AF768594

Birkhäuser

Progress in Mathematics
Volume 16

Series Editors
Hyman Bass
Joseph Oesterlé
Yuri Tschinkel
Alan Weinstein

For further volumes:
http://www.springer.com/series/4848

Phillip Griffiths • John Morgan

Rational Homotopy Theory and Differential Forms

Second Edition

Birkhäuser

Phillip Griffiths
Institute for Advanced Study
Princeton University
Princeton, NJ, USA

John Morgan
Simons Center for Geometry and Physics
Stony Brook University
Stony Brook, NY, USA

ISBN 978-1-4939-3699-1 ISBN 978-1-4614-8468-4 (eBook)
DOI 10.1007/978-1-4614-8468-4
Springer New York Heidelberg Dordrecht London

Mathematics Subject Classification (2010): 57R20, 57R57, 58A10, 58A14, 55R20, 55R10, 55R45, 55R40

© Springer Science+Business Media New York 2013
Softcover reprint of the hardcover 2nd edition 2013
This work is subject to copyright. All rights are reserved by the Publisher, whether the whole or part of the material is concerned, specifically the rights of translation, reprinting, reuse of illustrations, recitation, broadcasting, reproduction on microfilms or in any other physical way, and transmission or information storage and retrieval, electronic adaptation, computer software, or by similar or dissimilar methodology now known or hereafter developed. Exempted from this legal reservation are brief excerpts in connection with reviews or scholarly analysis or material supplied specifically for the purpose of being entered and executed on a computer system, for exclusive use by the purchaser of the work. Duplication of this publication or parts thereof is permitted only under the provisions of the Copyright Law of the Publisher's location, in its current version, and permission for use must always be obtained from Springer. Permissions for use may be obtained through RightsLink at the Copyright Clearance Center. Violations are liable to prosecution under the respective Copyright Law.
The use of general descriptive names, registered names, trademarks, service marks, etc. in this publication does not imply, even in the absence of a specific statement, that such names are exempt from the relevant protective laws and regulations and therefore free for general use.
While the advice and information in this book are believed to be true and accurate at the date of publication, neither the authors nor the editors nor the publisher can accept any legal responsibility for any errors or omissions that may be made. The publisher makes no warranty, express or implied, with respect to the material contained herein.

Printed on acid-free paper

Springer is part of Springer Science+Business Media (www.birkhauser-science.com)

Preface to the First Edition

This monograph originated as a set of informal notes from a summer course taught by the present authors, together with Eric Friedlander, at the Istituto Matematico "Ulisse Dini" in Florence during the summer of 1972. Even though more formal expositions of Sullivan's theory have since appeared, including the major original source [26], there has been a steady continuing demand for the old Florence notes. Moreover, one of us (J.M.) has become involved in the subject again through a series of lectures given at the University of Utah in January, 1980, together with joint work in progress with James Carlson and Herb Clemens on a new type of application of the theory to algebraic geometry. Since the Florence notes represented an approach and point of view that does not appear in the literature, we decided to publish the present revised and corrected version.

The material in this monograph is outlined in the table of contents and is informally discussed in the introduction below. Here we should like to observe that the text roughly divides into two parts. The first seven chapters essentially constitute an introductory course in algebraic topology with emphasis on homotopy theory. The main prerequisite is some familiarity with simplicial homology, covering spaces, and CW complexes.

Chapters 9–15 cover the main topic of differential forms and homotopy theory, with emphasis on the homotopy-theoretic and functorial properties of differential graded algebras and minimal models, a topic that does not appear explicitly in detail in the literature. An extensive set of exercises, frequently with copious hints, forms an essential complement to the material in the text.

We would like to make several acknowledgements to colleagues whose help and advice have been invaluable. The first and foremost is to Dennis Sullivan. It was he who introduced us to the idea of relating homotopy theory and differential forms and who explained to us his theory around which these notes are built.

The second is to Francesco Gherardelli who organized the original summer course and to the Istituto Matematico "Ulisse Dini" and the city of Florence, which together provided excellent mathematical and cultural conditions for the initial preparation of the notes. While in Florence we benefited from conversations with Ngo Van Que, Jim Carlson, and Mark Green. Finally, Moishe Breiner prepared a beautifully handwritten set of notes that constituted the original version of this monograph.

We would also like to thank the University of Utah and abovementioned coworkers of J.M. for providing support and motivation leading to the revision of the Florence notes.

Finally we would like to point out two predecessors of the present theory. The first is Whitney's book [27]. As explained to us by Sullivan, this book contains the genesis of the use of differential forms to solve the commutative cochain problem and thus get the homotopy type of the space. The main thing lacking at the time Whitney wrote the book was the $\mathbb{Q}$-structure. Secondly the relationship between differential forms and homotopy theory was anticipated by Chen [2].

Many of the results we find from a general viewpoint were established, frequently in stronger form, by him using the method of iterated integrals.

Preface to the Second Edition

Thirty years have passed since the publication of the first edition, and we felt this monograph deserved updating. The essential structure and presentation remains the same, but several major additions seemed appropriate. We have included in an appendix a proof of the correspondence between rational minimal models and rational Postnikov towers different, and we feel more intuitive, from the one presented in the first edition. This proof relies on a set of polynomial forms with a filtration whose Serre spectral sequence agrees with the usual one for a fibration. We have also added a chapter describing Quillen's approach to rational homotopy theory and comparing and contrasting it with Sullivan's. Lastly, we have added a chapter on operads and A_∞ algebras and indicated briefly how the commutative version, C_∞-algebras, gives another algebraic description of rational homotopy theory.

The second author thanks Mohammed Abouzaid and Bruno Vallette for their help with the chapter on A_∞-structures and Eric Malm for his help in preparing the diagrams and figures.

Princeton, NJ, USA — Phillip A. Griffiths
Stony Brook, NY, USA — John W. Morgan

Contents

Chapter 1
Introduction

The purpose of this course is to relate the C^∞-differential forms on a manifold to algebro-topological invariants. A model of results along these lines is deRham's theorem, which says that the cohomology of the differential graded algebra (DGA) of C^∞-forms is isomorphic to the singular cohomology with coefficients in $\mathbb{R}$, i.e.

$$H^*_{dR}(M) \cong H^*(M, \mathbb{R}) \quad (C^\infty \text{ deRham theorem}).$$

The main theorem of this course will be that from the DGA of C^∞-forms, it is possible to calculate all of the real algebro-topological invariants of the manifold. More precisely, we shall be able to use the forms to obtain the (Postnikov tower) tensored with $\mathbb{R}$ of the manifold.

In the next seven chapters of this book, we shall discuss the standard terminology, objects, and theorems of elementary homotopy theory, culminating in the description of the Postnikov tower of a space. We then define the *localization* of a CW complex at 0; this allows us to take a CW complex and replace it by one in which all torsion and divisibility phenomena have been removed (allowing one to focus on the $\mathbb{Q}$-information in the original space). When we compare the Postnikov tower of the original space with that of its localization, we see that all the relevant information (homotopy and homology groups, k-invariants) has been tensored with $\mathbb{Q}$.

Once we have established these basic facts, we turn to the main theorem as shown to us by Sullivan. First, we define the rational p.l. forms on a simplicial complex K. They form a DGA defined over $\mathbb{Q}$. By integration, these forms give $\mathbb{Q}$-valued simplicial cochains on K, and this integration process induces an isomorphism of the cohomology of the rational p.l. forms to the usual (simplicial or singular) rational cohomology of the space:

$$H^*_{p.l.}(K) \cong H^*(K, \mathbb{Q}) \quad (\text{p.l. deRham theorem}).$$

There are two very important points here. The first is that we are working over $\mathbb{Q}$ rather than $\mathbb{R}$, as we would be forced to do with C^∞-forms. The second is that the

P. Griffiths and J. Morgan, *Rational Homotopy Theory and Differential Forms*, Progress in Mathematics 16, DOI 10.1007/978-1-4614-8468-4_1,

© Springer Science+Business Media New York 2013

p.l. forms are a differential, graded-commutative algebra—the simplicial or singular cochains over $\mathbb{Q}$ are not commutative. Thus, the p.l. forms have a good property of ordinary cochains (they are defined over $\mathbb{Q}$) and a good property of C^∞-forms (they are graded commutative). Both these properties are essential.

Next, we turn to the homotopy theory of DGAs, which are always implicitly assumed to be associative and graded commutative. Given one such, $\mathcal{A}$, we show how to extract a *minimal model* for it. This is a DGA, $\mathfrak{M}_{\mathcal{A}}$, which satisfies some internal condition, together with a map of DGAs:

$$\rho_{\mathcal{A}}: \mathfrak{M}_{\mathcal{A}} \longrightarrow \mathcal{A}$$

which induces an isomorphism on cohomology.

In the case that $H^1(\mathcal{A}) = 0$, the internal properties that $\mathfrak{M}_{\mathcal{A}}$ is required to satisfy are:

1. It is free as a graded-commutative algebra with generators in degrees ≥ 2 only.
2. For all $x \in \mathfrak{M}_{\mathcal{A}}$, the element dx is decomposable.

It turns out that, given $\mathcal{A}$, these properties characterize $\mathfrak{M}_{\mathcal{A}}$ up to isomorphism.

We shall show, in addition, that when $\mathcal{A}$ is the algebra of p.l. forms on a simply connected, simplicial complex X, then $\mathfrak{M}_{\mathcal{A}}$ is dual to the rational Postnikov tower of X.

The duality between minimal models defined over $\mathbb{Q}$ and rational Postnikov towers is described in Chap. 12. Schematically we have a "commutative diagram":

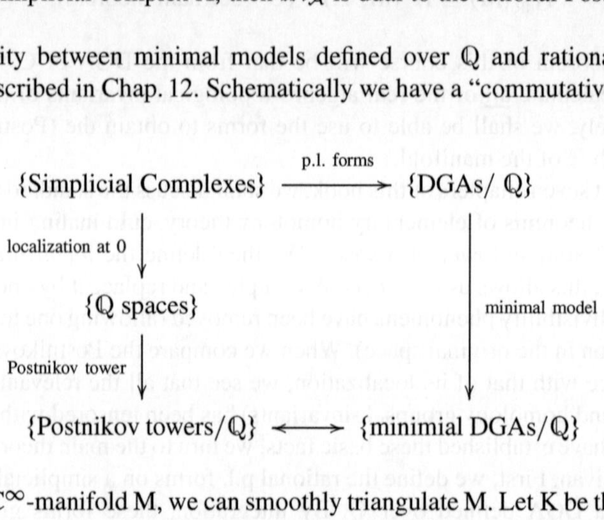

Given a C^∞-manifold M, we can smoothly triangulate M. Let K be the simplicial complex of this triangulation. We have both the C^∞-forms on M and p.l. forms on K. These are both included in $A^*_{p.C^\infty}(M)$ the DGA of "piecewise C^∞-forms" on M, (i.e., the forms whose restriction to each simplex of the triangulation is smooth), and the inclusions

$$A^*_{C^\infty}(M) \longrightarrow A^*_{p.\,C^\infty}(M) \longleftarrow A^*_{p.l.}(K) \otimes_{\mathbb{Q}} \mathbb{R}$$

induce isomorphisms in cohomology. From this comparison theorem, it follows that the minimal models satisfy

$$\mathfrak{M}(A^*_{C^\infty}(M)) \cong \mathfrak{M}(A^*_{p.l.}(M)) \otimes_{\mathbb{Q}} \mathbb{R}.$$

This is the precise statement that "the deRham complex contains all the real algebraic-topological information from the manifold M." Schematically the theory is arranged as follows:

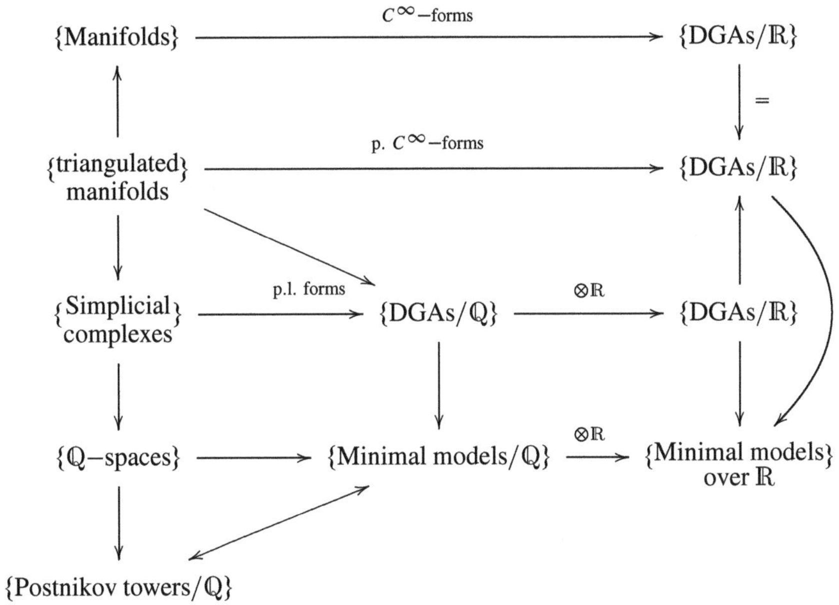

Though these notes concentrate mainly on the case of simply connected spaces, there are generalizations to the nonsimply connected case. In purely algebraic terms, part of the theory of the nonsimply connected case is similar to the simply connected one. When we try to make comparisons with homotopy, the results are much weaker. The available information from the algebra of forms which is most meaningful in classical terms deals with the fundamental group. Chapter 13 discusses this.

Chapter 2
Basic Concepts

Here we give a brief introduction to the basics of CW complexes, homotopy theory, homology, and the algebraic topology of manifolds. Here are some more references for more details on the material in this chapter. For a good introduction to CW complexes, homology, and cohomology, consult Greenberg's book [7]. For a more encyclopedic treatise on algebraic topology which covers all the homotopy theory presented in this course, save localization, one should see Spanier's book [23]. For another account of some of the topics presented later in this course, such as obstruction theory, one should see Hu's book [9].

2.1 CW Complexes

It will suffice for the purposes of this course (and for most other situations, also) to do homotopy theory for a restricted class of spaces. These are the spaces which are homotopy equivalent to CW complexes. All naturally encountered spaces have this property (e.g., manifolds, algebraic varieties, loop spaces on CW complexes, $K(\pi, n)$s). Moreover for these spaces, the Whitehead theorem which states that ($f: X \to Y$ is a homotopy equivalence if and only if f_* is an isomorphism on homotopy groups—cf. Sect. 9.2 for a proof) is true. What this means is that the usual functors of homotopy theory are powerful enough to decide when two CW complexes are homotopically equivalent.

We begin with the definition of a CW complex. Let D^n denote the unit n-disk, namely,

$$D^n = \{x = (x_1, \ldots, x_n) \in \mathbb{R}^n : ||x||^2 \leq 1\}$$

and let S^{n-1} denote the unit $(n-1)$ sphere, i.e., the boundary ∂D^n of D^n. (Note: In these notes, we have also used the notation e^n, for n-cell, as another symbol for D^n.) Given X and a continuous map $f: S^{n-1} \to X$, we form the *adjunction space*

P. Griffiths and J. Morgan, *Rational Homotopy Theory and Differential Forms*, Progress in Mathematics 16, DOI 10.1007/978-1-4614-8468-4_2,
© Springer Science+Business Media New York 2013

$$X \cup_f D^n$$

which is the quotient space of the disjoint union of X and D^n where every $a \in \partial D^n$ is identified with $f(a) \in X$. (Note: Above we required that f be a continuous map; usually we shall omit mention of continuity, with the understanding that map means continuous function.) Geometrically, what we have done is attach an n-cell to X

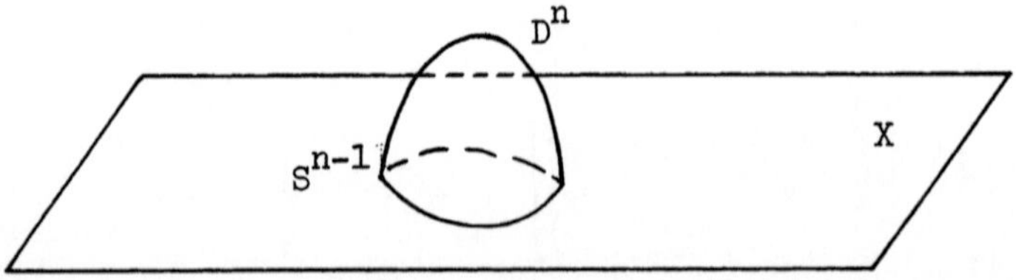

To give a space X, the structure of a CW complex means intuitively that X is obtained from a point by successively attaching cells. More precisely we have subspaces $X^{(i)}$ of X with

$$\phi = X^{(-1)} \subset X^{(0)} \subset X^{(1)} \subset \ldots, \quad X = \cup_{i=0}^{\infty} X^{(i)}$$

such that (1) $X^{(i+1)}$ is obtained from $X^{(i)}$ by attaching (i + 1)-cells and (2) if $X \neq X^{(n)}$ for any n (thus X is infinite dimensional), then X has the *weak topology* with respect to the $X^{(n)}$'s meaning that $U \subset X$ is an open set if and only if $U \cap X^{(n)}$ is open for all n. We call $X^{(n)}$ the n-skeleton of X. (Note: Infinite-dimensional CW complexes such as $\mathbb{C}P^{\infty}$, the infinite Grassmannians, the infinite sphere, etc. are very useful in homotopy theory. The weak topology means that in all cases, "∞" can be well approximated by "arbitrarily large n." Thus, for example, a map of a compact space into the infinite CW complex X is simply given by a map into X^N for large some N. Also, a map $f: X \to Y$ from a CW complex to a space is continuous if and only if its restriction to each skeleton $X^{(n)}$ is continuous.

Examples of CW Complexes

1. The n-sphere $S^n = \{pt.\} \cup_f D^n$ where $f: \partial D^n \to \{pt.\}$ is a degenerate attaching map.
2. The *complex projective space* $\mathbb{C}P^n$ is given a CW structure inductively by

$$\mathbb{C}P^n = \mathbb{C}P^{n-1} \cup_f D^{2n}$$

where

$$f: S^{2n-1} \longrightarrow \mathbb{C}P^{n-1}$$

is the *Hopf map*. More precisely, if we think of $\mathbb{C}P^n$ as the lines through the origin in $\mathbb{C}^{n+1}$, then taking D^{2n} to be the unit ball in $\mathbb{C}^n$, the attaching map $f\colon \partial D^{2n} \to \mathbb{C}P^{n-1}$ assigns to each point on the unit sphere in $\mathbb{C}^n$ the complex line joining through the origin containing that point 5.

3. $\mathbb{C}P^\infty = \lim_{n\to\infty}\mathbb{C}P^n$ is the infinite CW complex having one 2n-cell for each $n \in \mathbb{Z}^+$ and with the attaching maps given as above.
4. Any *simplicial complex* K has the natural structure of a CW complex. The n-cells of this CW structure are exactly the n-simplices. Conversely, if X is a CW complex, then there is a simplicial complex K and a homotopy equivalence from K to X (cf., Exercise (13)).
5. A *CW pair* (X, A) is a pair of spaces $A \subset X$ such that X is obtained from A by attaching cells. (It is not necessary that A itself be a CW complex.) If (X, A) is a CW pair, then we denote by $X^{(n)} \cup A$ the union of A with all cells of dimension $\leq$ n. Again, if X is obtained by attaching infinitely many cells to A, then X is given the limit (or weak) topology. If X is a CW complex and $A \subset X$ is a subcomplex, then (X, A) is a CW pair.

CW complexes are constructed so that, almost by definition, one works inductively up through the skeleton. As an example of this, we prove the *homotopy extension theorem* for CW pairs.

Theorem 2.1. *Given a CW pair* (Y, X)*, a map* $f\colon Y \to Z$*, and a homotopy* $F\colon X \times I \to Z$ *with* $F|_{X\times\{0\}} = f|_X$*, then there is an extension* $G\colon Y \times I \to Z$ *of* F *such that* $G(y, 0) = f(y)$.

Proof.

Step I: Given $f\colon D^n \to Z$ and $F\colon S^{n-1} \times I \to Z$ with $F|_{S^{n-1}\times\{0\}} = f|_{\partial D^n}$, find $G\colon D \times I \to Z$ extending F with $G|_{D^n\times\{0\}} = f$.

In the picture of $D^n \times I$

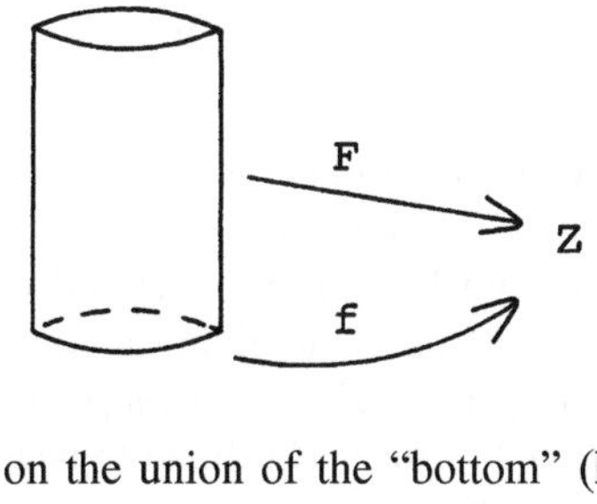

we are given a map G_0 on the union of the "bottom" ($D^n \times \{0\}$) and the "side" ($\partial D^n \times I$). We want to extend to a map on all of $D^n \times I$.

This is done taking the projection p of $D^n \times I$ onto $(D^n \times \{0\}) \cup (S^{n-1} \times I)$ from the point $\{(\text{middle of } D^n) \times \{2\}\}$

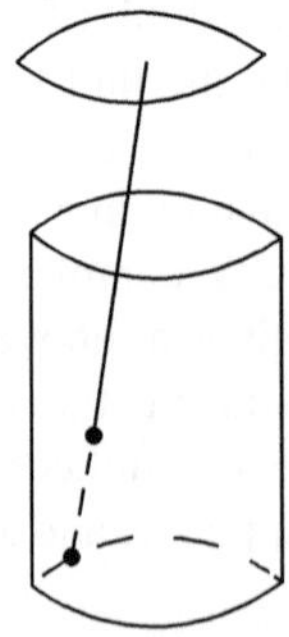

and defining $G(y,t) = G_0(p(y,t))$. Note: In this argument, as throughout homotopy theory, 99% of the proof is to find the correct "picture." If this is done properly, no geometric argument will be difficult (although some algebraic computations may be messy).

Step II: Given $f\colon Y \to Z$ and $F\colon X \times I \to Z$, we shall inductively construct $G^{(i)}\colon (Y \times \{0\}) \cup [(X \cup Y^{(i)}) \times I\} \to Z$. Given $G^{(i-1)}$, consider any i-cell D^i_α and attaching map $\partial D^i_a \to Y^{(i-1)}$. Then we have

$$G^{(i-1)} \circ f_\alpha \times I\colon (S^{i-1} \times I) \cup (D^i_\alpha \times \{0\}) \longrightarrow Z$$

and we may use Step I to extend this map over $D^i_\alpha \times 1$ to a map $G^{(i)}_a$. Doing this over each i-cell and taking the union of the maps give $G^{(i)}$. Let $G = \cup_i G^{(i)}$ (i.e., $G|Y^{(i)} = G^{(i)}$). By the definition of the weak topology, G is continuous and gives the required extension of F, by the construction. ■

2.2 First Notions from Homotopy Theory

In homotopy theory, one always considers CW complexes modulo an equivalence relation, that of *homotopy equivalence.* Two maps $f_0, f_1\colon X \to Y$ are *homotopic* if there exists $F\colon X \times I \to Y$ with $F(x,0) = f_0(x)$ and $F(x,1) = f_1(x)$. We may set $f_t(x) = F(z,t)$ and think of the f_t as giving a continuous deformation of f_0 into f_1. Maps $f\colon X \to Y$ and $g\colon Y \to X$ are *homotopy inverses* if $g \circ f \sim id_X$ and $f \circ g \sim id_Y$ (here, the notation "$\sim$" means "is homotopic to," and $id_X\colon X \to X$ is the identity map). A map $f\colon X \to Y$ is a *homotopy equivalence* if it has a homotopy inverse; X and Y are *homotopy equivalent* if there is a homotopy equivalence $f\colon X \to Y$. Homotopy equivalence is an equivalence relation on the collection of CW complexes. In homotopy theory, one considers spaces as equivalent if they are homotopy equivalent (in particular, topological dimension is not defined). An equivalence class of homotopy equivalent spaces is said to be a *homotopy type.* For the rest of this course, space will mean CW complex, with the only exception that we shall speak of path spaces and loop spaces on CW complexes (definitions below), which are not CW complexes as they stand. However, they always have the homotopy type of a CW complex [17] and so may be unambiguously considered as "spaces."

While the category of CW complexes is quite flexible and easy to work with for many applications in homotopy theory, sometimes, for example, in defining an appropriate DGA of forms, it is better to work with simplicial complexes.Of course, as we have already remarked, any simplicial complex is a CW complex. The converse is true up to homotopy.

Lemma 2.2. *Let* X *be a CW complex. Then there is a simplicial complex* K *homotopy equivalent to* X.

Proof. The proof is by induction over the skeleta of X. Suppose inductively that for some $n \geq 0$, we have a simplicial complex K^n and a homotopy equivalence $\varphi_n\colon K^n \to X^{(n)}$. For each $(n+1)$-cell e_α of X, we have the attaching map $f_\alpha\colon S^n \to X^n$. There is a map $g_\alpha\colon \partial\Delta^{n+1} \to K^n$ with $\varphi_n \circ g_\alpha$ homotopic to f_α. By subdividing the standard triangulation of $\partial\Delta^{n+1}$ to produce a triangulation τ of $\partial\Delta^{n+1}$, we can arrange that g_α is a simplicial map (without subdividing K^n). Consider a collar neighborhood C of $\partial\Delta^{n+1}$ in Δ^{n+1} with $\partial\Delta^{n+1}$ corresponding to the 0-end. The product $\tau \times I$ defines a linear cell structure on C. We form the quotient $\bar{C}$ of C where we identify points in $\partial\Delta^{n+1} \subset C$ if they have the same image under g_α. Then the image in $\bar{C}$ of the product of each simplex of τ with I is a cell. Let B be the image of $\partial\Delta^{n+1}$ in $\bar{C}$. Now by induction on the simplices σ of the triangulation τ, we triangulate the image $\bar{\sigma}$ of $\sigma \times I$ in $\bar{C}$ without subdividing $\bar{\sigma} \cap B$. Suppose that we have done this for $(\overline{\partial\sigma})$. At the 1-end, take the cone to the barycenter of $\sigma \times \{1\}$ of the triangulation of $\partial\sigma \times \{1\}$. This produces a triangulation of the boundary of the cell $\bar{\sigma}$ in $\bar{C}$, a triangulation that does not subdivide the image of $\sigma \times \{0\}$ in $\bar{C}$. We then take the cone over this triangulation to $v_\sigma \times \{1/2\}$, where v_σ is the barycenter of σ. Then we extend the triangulation on the 1-end of C to a triangulation over the rest of Δ^{n+1}. The produces a simplicial complex K', containing K as a subcomplex, and there is an extension of the homotopy equivalence from K' to $X^{(n)} \cup_{f_\alpha} e_\alpha$ extending φ_n. Performing this construction simultaneously for every $(n+1)$-cell of X produces a simplicial complex K^{n+1} containing K^n and a homotopy equivalence $\varphi_{n+1}\colon K^{n+1} \to X^{(n)}$ extending φ_n. Taking the limit (i.e., increasing union) over all n establishes the result. ∎

In many problems in homotopy theory, we wish to make a construction relative to a map

$$f\colon X \longrightarrow Y.$$

It is frequently easier to work with an inclusion rather than an arbitrary map. This is always possible up to homotopy equivalence.

Theorem 2.3. *Given* $f\colon X \to Y$, *there is a space* M_f, *the mapping cylinder of* f, *inclusions* $j\colon X \to M_f$ *and* $i\colon Y \to M_f$, *and a map* $\pi\colon M_f \to Y$

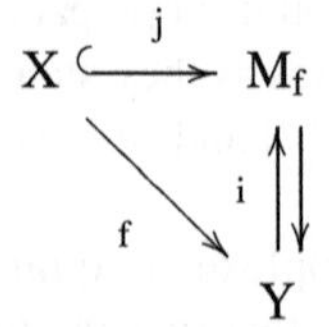

where π *and* i *are homotopy inverses and* $\pi \circ j = f$. *Thus, we may replace* Y *by a homotopy equivalent space in which X is included.)*

Proof. Define $M_f = (X \times I) \cup_f Y$

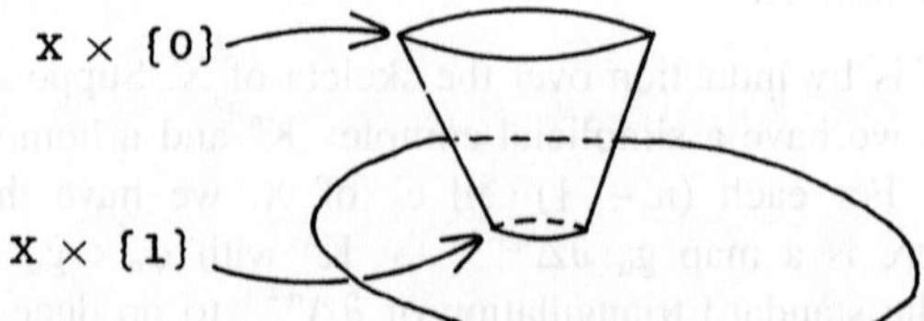

where (x, 1) is identified with $f(x) \in Y$. Then $\pi\colon M_f \to Y$ is given by $\pi(x,t) = f(x)$ for all $x \in X$ and $\pi(y) = y$ for all $y \in Y$ (this is consistent), and this gives a retraction of M_f onto Y. ■

Note: If X and Y are CW complexes and $f\colon X \to Y$ is a *cellular map* (i.e., $f(X^{(i)}) \subset Y^{(i)}$), it is easy to give M_f the structure of a CW complex. In Exercise (32), a proof of the fact that any f is homotopic to a cellular map f′ is outlined, so that $M_f \sim M_{f'}$ which is a CW complex (the notation $A \sim B$ means that A and B are homotopy equivalent). Thus, we may consider any map as an inclusion without leaving the category of CW complexes.

Above we discussed the homotopy extension property (h. e. p.) and proved that a subcomplex of a CW complex always has the h. e. p. Now there is a dual property to the h. e. p. called the *homotopy lifting property* or *covering homotopy property*. Given spaces E, B, we say that $\pi\colon E \to B$ has the *homotopy lifting property* if given any space Y, a map $Y \xrightarrow{f} E$, and a homotopy g_t of $g = \pi \circ f$, there is a homotopy f_t of f such that $\pi \circ f_t = g_t$ (thus, the homotopy f_t "covers" or "lifts" the homotopy g_t).

Here f is said to be a *lifting* of g, and covering homotopy says that if a map g can be lifted, then any homotopy g_t of g can be lifted also. Not all maps have the homotopy lifting property; e.g., if B is connected, then π must be onto. If $\pi\colon E \to B$ has the h. l. p. (= homotopy lifting property), then it is said to be a *fibration*. For any $b \in B$, the fiber $F_b = \pi^{-1}(b)$ is the preimage of the point. In a fibration, any two fibers are homotopy equivalent provided that the base is path connected (cf. Chap. 4). We let F be any space having the homotopy type of F_b (F is called a typical fiber). We write the fibration as

$$\begin{array}{ccc} F & \longrightarrow & E \\ & & \downarrow \pi \\ & & B \end{array}$$

having in mind a picture like

$$\begin{array}{ccc} F_b & \longrightarrow & E \\ \downarrow & & \downarrow \\ \{b\} & \longrightarrow & B \end{array}$$

Examples of Fibrations

1. Locally trivial *fiber bundles*, *vector bundles*, and the associated *sphere bundles*, *covering spaces*, are all examples of fibrations (for a discussion of these, cf. [25]).
2. Let X be a space with a base point x_0. Define the *path space* based at $x_0 \in X$, $\mathcal{P}(X, x_0)$, to be the set of all paths given by maps $\omega: I \to X$, $\omega(0) = x_0$. The topology on $\mathcal{P}(X)$ is the compact-open topology. Thus, a sub-basis for the open sets in $\mathcal{P}(X)$ is given by taking $K \subset I$ a compact subset and $U \subset X$ an open set and letting $< K, U >$ be all maps $\omega: I \to X$ with $\omega(K) \subset U$. Define $\pi: \mathcal{P}(X) \to X$ by $\pi(\omega) = \omega(1)$. This is a fibration.

Homotopy Exact Sequence of a Fibration. Here is the statement:

Theorem 2.4. *Suppose that* $\pi: E \to B$ *is a fibration with* B *path connected. Fix* $b \in B$, *and let* F_b *be the fiber over* b *and* $i: F_b \to E$ *the inclusion. Finally, fix* $e \in F_b$. *Then we have a long exact sequence of homotopy groups:*

$$\longrightarrow \pi_n(F_b, e) \xrightarrow{i_\#} \pi_n(E, e) \xrightarrow{\pi_\#} \pi_n(B, b) \longrightarrow \pi_{n-1}(F_b, e) \longrightarrow .$$

Proof.
Exactness at $\pi_n(E, e)$. It is clear that $\pi_\# \circ i_\# = 0$. Suppose $a \in \pi_n(E, e)$ and $\pi_\#(a) = 0$. Represent a by a map $\varphi: (S^n, p) \to (E, e)$, where $p \in S^n$ is the base point. Since $\pi_\#(a) = 0$, there is a homotopy H from $\pi \circ \varphi: (S^n, p) \to (B, b)$ to the constant map at b, a homotopy that is constant on $\{p\} \times I$. Use the homotopy lifting property for the relative CW complex (S^n, p) to lift H to a homotopy $\tilde{H}: S^n \times I \to E$ beginning at φ and sending $\{p\} \times I$ to e. The map $H|_{S^n \times \{1\}}$ is a map $(S^n, p) \to (F_b, e)$ representing $a \in \pi_n(E, e)$. This proves that the image of $i_\#$ contains the kernel of $\pi_\#$, completing the proof of exactness at $\pi_n(E, e)$.
Exactness at $\pi_n(B, b)$. Let us first define the connecting homomorphism $\pi_n(B, b) \to \pi_{n-1}(F_b, e)$. Given an element, $\bar{a} \in \pi_n(B, b)$ represent it by a map $\psi: (S^n, p) \to (B, b)$. Let $p: (D^n, \partial D^n) \to (S^n, p)$ be the map collapsing the boundary

of the disk to the base point of the sphere. Use the fact that the disk is contractible and the homotopy lifting property to lift the composite of p followed by ψ to a map $\tilde{\psi}: (D^n, \partial D^n) \to (E, F_b)$ sending the base point (in ∂D^n) to e. Then the restriction of $\tilde{\psi}|_{\partial D^n}$ represents the image under the connecting homomorphism of a. We leave the proof that this is process determines well-defined homomorphism to the reader. It is clear from this construction that the composition of $\pi_\#$ followed by the connecting homomorphism is zero since we can take the lift of $\bar{a} = \pi_\#(a)$ to be the composition the collapsing map $p: (D^n, \partial D^n) \to (S^n, p)$ followed by a. It follows that the image of the connecting homomorphism applied to $\pi_\#(a) = 0$. Conversely, if the image of the connecting homomorphism applied to $\bar{a}$ is zero, then the lifted map $(D^n, \partial D^n) \to (E, F_b)$ has the property that its restriction to the boundary is homotopic in (F_b, e) to a point map. Homotopy extension allows us to lift the original map of S^n to B to a map of S^n to E, showing that the image of $\pi_\#$ contains the kernel of the connecting homomorphism.
Exactness at $\pi_n(F_b, e)$. From the construction, any based sphere in (F_b, e) coming from the connecting homomorphism bounds a disk in (E, e), showing that the composition of the connecting homomorphism followed by $i_\#$ is zero. Conversely, if an element $c \in \pi_n(F_b, e)$ is trivial in $\pi_n(E, e)$, then the sphere in (F_b, e) representing c bounds a disk in (E, e) whose image under $\pi_\#$ is a sphere of one higher dimension in (B, b) whose image under the connecting homomorphism is c. ■

The Loop Space

Proposition 2.5. *Let* $x_0 \in X$ *be given and let* $\mathcal{P}(X, x_0)$ *denote the space of paths in* X *beginning at* x_0. *Then* $\pi: \mathcal{P}(X, x_0) \to X$ *is a fibration.*

Proof. Given a path $g: I \to X$ and an element $\tilde{g}_0 \in \mathcal{P}(X)$ such that $\pi(\tilde{g}_0) = g(0)$ (i.e., given a path g in X and a path g_0 beginning at x_0 and ending at g(0)), we define $\tilde{g}_t \in \mathcal{P}(X, x_0)$ by

$$\tilde{g}_t(s) = \begin{cases} \tilde{g}_0(s(1+t)) & 0 \le s \le \frac{1}{1+t} \\ g(s(1+t)-1) & \frac{1}{1+t} \le s \le 1. \end{cases}$$

One sees easily that $\pi(\tilde{g}_t) = g_t$ and that $t \to \tilde{g}_t$ is a continuous mapping of I into $\mathcal{P}(X, x_0)$. This proves the homotopy lifting property for points. One checks that the construction varies continuously with the original data and hence gives the homotopy lifting property for all spaces. ■

Definition. The fiber $\pi^{-1}(x_0) \subset \mathcal{P}(X, x_0)$ is denoted ˙(X, x_0) and is the *loop space of* X *based at* x_0.

Lemma 2.6. *$\mathcal{P}(X, x_0)$ is contractible (i.e., homotopy equivalent to a point).*

Proof. Define $\mathcal{P}(X, x_0) \times I \xrightarrow{c} \mathcal{P}(X, x_0)$ by

$$c(\omega, t)(s) = \omega(ts).$$

Clearly, $c(\omega, 0)$ is the constant path at x_0 for any $\omega \in \mathcal{P}(X, x_0)$, and $c(\omega, 1) = \omega$. ■

Corollary 2.7. *Fix a point $x_0 \in X$ and let ω_0 be the constant loop at x_0. There is an isomorphism $\pi_{n-1}(\dot{}(X, x_0), \omega_0) \xrightarrow{\cong} \pi_n(X, x_0)$.*

Proof. We have the fibration $\mathcal{P}(X, x_0) \to X$ with fiber $\dot{}(X, x_0)$ over the base point $x_0 \in X$. The result now follows from Theorem 2.4 and the fact that $\mathcal{P}(X, x_0)$ is contractible and hence has trivial homotopy groups. ■

We have seen how to replace any map by an inclusion up to homotopy. It is also possible to replace any map by a fibration up to homotopy. Given $f: X \to Y$, form the set of pairs $\{(x, \omega)\}$ where $x \in X$ and ω is a path in Y satisfying $\omega(0) = f(x)\}$. This is a subspace of $X \times \tilde{\mathcal{P}}(Y)$ where $\tilde{\mathcal{P}}(Y)$ represents the space of <u>all</u> paths in Y. (Thus, $\tilde{\mathcal{P}}(Y) = Y^I$ with the compact-open topology.) Call this subspace $\tilde{X}$. Define $\pi: \tilde{X} \to Y$ by $\pi(x, \omega) = \omega(1)$. Define $i: X \hookrightarrow \tilde{X}$ by $i(x) =$(x, constant path at $f(x)$). One checks easily that $\pi \circ i = f$, that $i: X \hookrightarrow \tilde{X}$ is a homotopy equivalence, and that $\pi: \tilde{X} \to Y$ is a fibration.

The freedom to convert maps up to homotopy into either inclusions or fibrations in homotopy category illustrates the elasticity of the notion of homotopy; to balance this flexibility, one has the perversity of nature that the homotopy groups of the simplest spaces, the spheres, have thus far been impossible to calculate.

2.3 Homology

A *chain complex* C_* is a set of abelian groups C_n indexed by the integers with maps, called *boundary maps*, $\partial_n: C_n \to C_{n-1}$ satisfying $\partial_{n-1} \circ \partial_n = 0$ for all n. The *homology* of a chain complex is a sequence of abelian groups defined by

$$H_n(C_*) = \text{Kernel}\, \partial_n / \text{Image}\, \partial_{n+1}.$$

The kernel of ∂_n is the group of *cycles* in degree n and the image of ∂_{n+1} is the group of *boundaries* in degree n so that the nth homology is the quotient of the group of cycles in degree n modulo the group of boundaries in degree n. Elements of $H_n(C_*)$ are said to have *degree* n.

Let X be a space. Here we shall introduce the construction of *singular homology* of X with $\mathbb{Z}$ coefficients and recall its basic properties.

Let Δ^n be the standard n-*simplex* in $\mathbb{R}^{n+1}$. This is the subset of $\mathbb{R}^{n+1}$ defined by

$$\Delta^n = \{(t_0, \ldots, t_n) \big| t_i \geq 0 \text{ for all } i \text{ and } \sum_{i=0}^{n} t_i = 1.\}$$

Let $\mathrm{Sing}_n(X)$ be the free abelian group generated by the *singular* n-*simplices*

$$\Delta^n \xrightarrow{f} X,$$

which by definition are the continuous maps of Δ^n to X. The ith-face of Δ^n, denoted f_i, is the subset where $t_i = 0$. It is naturally identified with Δ^{n-1} by an affine linear isomorphism that preserves the order of the vertices (the extremal points).

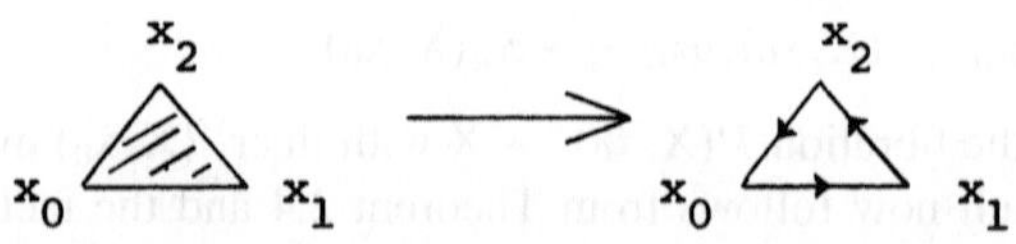

These identifications of the faces of Δ^n with Δ^{n-1} induce a boundary map

$$\partial_n\colon \mathrm{Sing}_n(X) \longrightarrow \mathrm{Sing}_{n-1}(X)$$

defined by

$$\partial_n \sigma = \sum_{i=0}^{n} (-1)^i \sigma|_{f_i}.$$

This boundary maps satisfy $\partial_{n-1} \circ \partial_n = 0$. This is the *singular chain complex* of a space:

$$\cdots \longrightarrow \mathrm{Sing}_{n+1}(X) \xrightarrow{\partial_{n+1}} \mathrm{Sing}_n(X) \xrightarrow{\partial_n} \mathrm{Sing}_{n-1}(X) \longrightarrow \cdots,$$

The *singular homology* of X, denoted $H_*(X)$, is the homology of the singular chain complex, i.e., it is defined as the quotient of the singular cycles modulo the singular boundaries; i.e.,

$$H_n(X) = \mathrm{Kernel}\partial_n / \partial_{n+1}\mathrm{Sing}_{n+1}(X)$$

Given a pair of spaces (X, A), we define the *relative singular chain complex* $\mathrm{Sing}_*(X, A)$ as the quotient $\mathrm{Sing}_*(X)/\mathrm{Sing}_*(A)$. This means the relative chain group indexed by n is the quotient $\mathrm{Sing}_n(X)/\mathrm{Sing}_n(A)$, and the boundary map is the one induced by the boundary map in the singular chain complex of X. The *relative singular homology* $H_*(X, A)$ is the homology of $\mathrm{Sing}_*(X, A)$. Notice that there is a natural map $H_n(X) \to H_n(X, A)$ induced by the quotient projection on the singular chain complexes. Also, there is a connecting homomorphism $H_n(X, A) \to H_{n-1}(A)$.

The basic properties of homology are:

(i) A map $f: X \to Y$ induces a map on homology

$$f_*: H_*(X) \longrightarrow H_*(Y)$$

which depends only on the homotopy class of f.

(ii) An orientation for S^n determines an isomorphism $H_n(S^n) \cong\to \mathbb{Z}$. If we reverse the orientation, then this isomorphism is multiplied by -1. A representative cycle for the class corresponding to $1 \in \mathbb{Z}$ is determined by a homeomorphism $\partial\Delta^{n+1} \to S^n$ which is orientation-preserving.

(iii) Given $Y \subset X$, there is an *exact homology sequence*

$$\ldots H_n(Y) \longrightarrow H_n(X) \longrightarrow H_n(X, Y) \xrightarrow{\partial} H_{n-1}(Y) \longrightarrow \ldots$$

(iv) If $U \subset Y \subset X$ is such that

$$\overline{U} \subset \text{interior } (Y),$$

then we have the *excision property*

$$H_*(X - U, Y - U) \cong H_*(X,\ Y).$$

(v) Any *homology theory*, that is, a functor from the homotopy category to abelian groups which satisfies (ii)–(iv), is naturally identified with singular homology [cf. Exercise (40)]. The most interesting of the axioms is probably excision, which is essentially the same as Mayer–Vietoris.

We define singular homology with coefficients in an abelian group G by forming the chain complex

$$\cdots \longrightarrow \mathrm{Sing}_n(X) \otimes G \xrightarrow{\partial_n \otimes 1} \mathrm{Sing}_{n-1}(X) \otimes G \longrightarrow \cdots$$

and taking its homology. These groups are denoted $H_n(X, G)$. When no coefficients are specified, $\mathbb{Z}$ coefficients are understood.

We also have *cohomology*. A cochain complex is a set of abelian groups $\{C^n\}$ indexed by the integers with coboundary maps $\delta^n: C^n \to C^{n+1}$ satisfying $\delta^{n+1} \circ \delta^n = 0$. The cohomology $H^n(C^*)$ is the quotient of kernel δ^n (the cocycles) modulo the image of δ^{n-1} (the coboundaries). The singular cohomology of X is the cohomology of the singular cochain complex

$$\xleftarrow{\delta_{n+1}} \mathrm{Hom}(\mathrm{Sing}_n(X), G) \xleftarrow{\delta_n} \mathrm{Hom}(\mathrm{Sing}_{n-1}(X), G) \longleftarrow \ldots$$

where δ^n is the map dual to ∂_{n+1}.

These groups are denoted $H^n(X, G)$. Again if no G is made explicit, $\mathbb{Z}$ coefficients are understood.

Products in Homology and Cohomology. Singular cohomology $H^n(X)$ has the structure of a *graded-commutative ring*. That is to say, there is a multiplication $H^p(X) \otimes H^q(X) \to H^{p+q}(X)$ with the property that $\alpha \cup \beta = (-1)^{\deg\alpha \cdot \deg\beta} \beta \cup \alpha$.

We begin with an explicit formula for the cup product in cohomology, *the Whitney formula*. Suppose that α is a singular cochain of degree p for X and β is a singular cochain of degree q for X. We define the cup product $\alpha \cup \beta$ by defining its value on any singular simplex $\sigma \colon \Delta^{p+q} \to X$ by

$$\langle \alpha \cup \beta, \sigma \rangle = \langle \alpha, \sigma|_{fr_p} \rangle \langle \beta, \sigma|_{bk_q} \rangle,$$

where fr_p is the p-dimensional face of Δ^{p+q}, the face spanned by the first $p+1$ vertices, and bk_q is the q-dimensional face of Δ^{p+q}, the face spanned by the last $q+1$ vertices. It is easy to see that this product is associative and satisfies

$$\delta(\alpha \cup \beta) = \delta(\alpha) \cup \beta + (-1)^p \alpha \cup \delta(\beta).$$

It follows easily that the cochain cup product induces an associative product on cohomology. The cup product has no symmetry properties on the cochain level; nevertheless, it turns out for classes $a \in H^p(X)$ and $b \in H^q(X)$, we have

$$a \cup b = (-1)^{pq} b \cup a.$$

That is to say, the induced product makes cohomology into an associative, graded-commutative ring with unit.

There are related algebraic structures:

(1) Homology has a co-associative, graded-commutative co-multiplication

$$\Delta \colon H_k(X) \to \oplus_{i+j=k} H_i(X) \otimes H_j(X).$$

(2) There is a cap product:

$$H^p(X) \otimes H_q(X) \to H_{q-p}(X).$$

The first pairing is induced by the following map on the chain level: For a singular simplex σ of degree k,

$$\sigma \mapsto \sum_{i=0}^{k} = \sigma|_{fr_i} \otimes \sigma|_{bk_{k-i}}.$$

The second pairing is induced by the following map on the chain level: Let σ be a singular simplex of degree q and let β be a singular cochain of degree p. Then

$$\sigma \cap \beta = \langle \beta, fr_p(\sigma) \rangle bk_{q-p}(\sigma).$$

Since singular homology and cohomology are defined using the same underlying chain complex, they are closely related, as expressed in the universal coefficient theorem. First, notice that singular n-cochains can be evaluated on singular n-chains that this leads to a pairing

$$H^n(X) \otimes H_n(X) \to \mathbb{Z},$$

and hence a homomorphism

$$H^n(X) \to \mathrm{Hom}(H_n(X), \mathbb{Z}).$$

This homomorphism appears in the universal coefficient theorem. First, we need to introduce Tor and Ext in the category of abelian groups. Suppose we are given two abelian groups A and B. There is a short free resolution of A, i.e., a short exact sequence

$$\{0\} \to F_1 \to F_0 \to A \to \{0\}$$

with F_0 and F_1 being free abelian groups. We define $\mathrm{Tor}(A, B)$ to be the kernel of the map $F_1 \otimes B \to F_0 \otimes B$, and we define $\mathrm{Ext}(A, rmB)$ to be the cokernel of the map $\mathrm{Hom}(F_0, B) \to \mathrm{Hom}(F_1, B)$. Up to canonical isomorphism, these definitions are independent of the choice of the free resolution of A.

Theorem 2.8 (Universal Coefficient Theorem). *1. There is a short exact sequence:*

$$0 \to \mathrm{Ext}(H_{n-1}(X), \mathbb{Z}) \to H^n(X) \to \mathrm{Hom}(H_n(X), \mathbb{Z}) \to 0.$$

2. For any abelian group G, *there are short exact sequences:*

$$\{0\} \to H_n(X) \otimes G \to H_n(X; G) \to \mathrm{Tor}(H_{n-1}(X), G) \to \{0\},$$

$$0 \to \mathrm{Ext}(H_{n-1}(X), G) \to H^n(X; G) \to \mathrm{Hom}(H_n(X), G) \to 0.$$

When the homology of X is finitely generated in each degree, this implies:

1. $H_n(X)/\mathrm{Torsion}$ and $H^n(X)/\mathrm{Torsion}$ are dual free abelian groups.
2. $\mathrm{Torsion}(H_{n-1}(X))$ and $\mathrm{Torsion}(H^n(X))$ are Pontrjagen dual finite abelian groups.

Properties of the Homology of Closed, Oriented Manifolds. The homology and cohomology groups of a closed, oriented manifold satisfy a strong relationship, Poincaré duality.

Theorem 2.9. *Let* M *be a closed, oriented manifold of dimension* n. *Then there is a fundamental class* $[M] \in H_n(M)$ *and cap product with this class induces an isomorphism*

$$\cap[M]: H^q(M) \to H_{n-q}(M).$$

While this theorem holds for topological manifolds, it has nice descriptions in the case when M is smooth. For a smooth manifold, the fundamental class $[M] \in H_n(M)$ can be described as follows. Take a smooth triangulation of M and choose a total ordering for the vertices of this triangulation. This induces an ordering of the vertices of each n-dimensional simplex and hence an identification of each n-dimensional simplex with Δ^n. For each n-dimensional simplex σ of the triangulation, we define a sign ϵ_σ by comparing the orientation on σ induced from that of M with the standard orientation of the n-simplex. Then the sum $c = \sum \epsilon_\sigma \sigma$, where the sum is taken over all the n-dimensional simplices of the triangulation, is a singular n-chain in M. Each $(n-1)$-dimensional simplex in the triangulation occurs as in the boundary of exactly two n-simplices, and it is easy to see that the signs cancel out so that c is a cycle. It represents the fundamental class of M.

Putting this together with the universal coefficient theorem, we see that Poincaré duality implies that the rank of $H_k(M)$ is equal to the rank of $H_{n-k}(M)$ and the torsion in $H_k(M)$ is isomorphic to the torsion in $H_{n-k-1}(M)$. Actually, there are much stronger statements than just abstract isomorphisms: There are dual pairings which we will now describe.

Given homology classes $a \in H_k(M)$ and $b \in H_{n-k}(M)$, we can represent them by singular cycles so that each singular simplex maps smoothly into M and furthermore so that the simplices in the representative for a and those for b meet transversely. This means that the intersection is a finite set of points and at each intersection point the tangent spaces of the two simplices are complementary subspaces in the tangent space of M. Since M and each of the simplices is oriented, we can associate a sign, ± 1, to each intersection point and the sum of these signs over all intersection points is an integer which depends only on the homology classes and not on the cycle representatives. This defines a pairing

$$H_k(M) \otimes H_{n-k}(M) \to \mathbb{Z}.$$

In fact, its adjoint is the composition of Poincaré duality followed by the usual evaluation map

$$H_k(M) \to H^{n-k}(M) \to \mathrm{Hom}(H_{n-k}(M), \mathbb{Z}),$$

so that the pairing is perfect on the quotients of these homology groups by their torsion subgroups. This is called the *intersection pairing*. The symmetry of cup product means that this pairing is also graded-commutative

$$a \cdot b = (-1)^{\deg(a) \cdot \deg(b)} b \cdot a.$$

There is an analogous pairing for the torsion groups: Given torsion classes $a \in H_k(M)$ and $b \in H_{n-k-1}(M)$, we choose representative cycles, α and β, that are disjoint. Then some multiple $N\alpha$ is the boundary of a $(k+1)$-chain ζ that we can make transverse to β. Counting signed intersections as before, we form $(\zeta \cdot \beta)/N \in \mathbb{Q}/\mathbb{Z}$. This defines a pairing

$$\text{Torsion } H_k(M) \otimes \text{Torsion } H_{n-k-1}(M) \to \mathbb{Q}/\mathbb{Z}$$

which is a perfect pairing in the sense that its adjoint is an isomorphism. This pairing is called the *linking pairing*.

This construction is most pleasing when the manifold is smooth and the cycle representatives are themselves closed, oriented smooth submanifolds of M. Notice that if P^k is a closed, oriented submanifold of M, then it represents a homology class $[P] \in H_k(M)$. One way to see this is to take the image of the fundamental class of P under the inclusion $P \hookrightarrow M$. Another is to take a smooth triangulation of P, choose a total ordering of the vertices of this triangulation, and then take the signed sum of the top-dimensional simplices forming a k cycle in M. Given closed, oriented submanifolds P^k and Q^{n-k} in M meeting transversally, we form the *homological intersection* $P \cdot Q$. This is the sum over the (finite) set of points of $P \cap Q$ of a sign for each point, a sign that compares on one hand the sum of the orientation for P at the intersection point followed by that of Q and on the other the orientation of M at that point. This homological intersection is exactly the value of the intersection pairing, as defined above using Poincaré duality, of the homology classes represented by P and Q.

2.4 Categories and Functors

Now we will introduce the language of *categories* and *functors* which so permeates mathematics today. There are two reasons for doing this. One is that it is encountered when reading reference articles, and the second is that it gives a convenient formalism for stating and remembering many of the results contained in these notes. In the notes, we do not always bother to give the category-theoretic reformulation of the results we prove. These will sometimes be left to the reader as exercises.

A *category*, $\mathcal{C}$, consists of a collection of objects, $\text{Obj}(\mathcal{C})$, and for any pair of objects A and B of the category a set of *morphisms* from A to B. This set is denoted by $\text{Hom}(A, B)$ or $\text{Hom}_{\mathcal{C}}(A, B)$. Several axioms are required to be satisfied:

1. $\text{Hom}(A, A)$ always contains a distinguished element I_A, the identity morphism of A.
2. For objects A, B, C of $\mathcal{C}$, there is a composition $\text{Hom}(A, B) \times \text{Hom}(B, C) \to \text{Hom}(A, C)$ which is associative $(fg)h = f(gh)$ and for which I_A is a unit; i.e., $f \circ I_A = f$ and $I_B \circ f = f$ for any $f \in \text{Hom}(A, B)$.

Note: The objects of $\mathcal{C}$ do not necessarily have to have underlying sets and a morphism from A to B does not have to be a set map. It is just considered as an arrow.

Examples. 1. **Abelian groups.** The objects are abelian groups; the morphisms are group homomorphisms.

2. **Topological spaces**. The objects are the usual topological spaces; the morphisms are continuous maps.
3. **Homotopy category**. The objects are spaces homotopy equivalent to CW complexes, and the morphisms are homotopy classes of maps.
4. Analogous to 1, we have the categories of groups, rings, fields, vector spaces over a field, and modules over a ring. An *isomorphism* in a category is a morphism with an inverse; i.e., $f: A \to B$ is an isomorphism if there exists $g: B \to A$ with $g \circ f = I_A$ and $f \circ g = I_B$.

The isomorphisms in 1, 2, and 3 are, respectively, group isomorphism, homeomorphism, and homotopy equivalence.

A *(covariant) functor* between two categories, $\mathcal{F}: \mathcal{C} \to \mathcal{C}'$, is an assignment of objects $\mathcal{F}(A)$ in $\mathcal{C}'$ to objects A in $\mathcal{C}$ and morphisms $\mathcal{F}(f)$ in $\mathrm{Hom}_{\mathcal{C}'}(\mathcal{F}(A), \mathcal{F}(B))$ to morphisms f in $\mathrm{Hom}_{\mathcal{C}}(A, B)$ such that identities and compositions are preserved. Clearly a functor sends isomorphic objects in $\mathcal{C}$ to isomorphic objects in $\mathcal{C}'$.

For any category $\mathcal{C}$, there is an *opposite category* $\mathcal{C}^{\mathrm{opp}}$. These two categories have the same objects and by definition $\mathrm{Hom}_{\mathcal{C}^{\mathrm{opp}}}(A, B) = \mathrm{Hom}_{\mathcal{C}}(B, A)$. A *contravariant* functor from $\mathcal{C}$ to $\mathcal{C}'$ is by definition a covariant functor from $\mathcal{C}^{\mathrm{opp}}$ to $\mathcal{C}'$.
A *homotopy functor* is a functor from the homotopy category to another category. Algebraic topology is the study of homotopy functors into algebraic categories—i.e., groups, rings, fields, chain complexes, and vector spaces. Under a homotopy functor, homotopy equivalent spaces have assigned to them isomorphic objects. This is why, in homotopy theory, it is permissible to identify homotopy equivalent objects.

As an example of a functor of algebraic topology, consider homology. The *homology functor* assigns to each space X a graded group $H_*(X)$ and to a continuous map $f: X \to Y$ an induced map on homology $f_*: H_*(X) \to H_*(Y)$. If $f \sim g$, then $f_* = g_*$ so that homology is actually a homotopy functor. (See the next chapter for more details on singular homology.)

Chapter 3
CW Homology Theorem

3.1 The Statement

Suppose that X is a CW complex and let $X^{(n)}$ denote its n-skeleton. Then the $(n+1)$-skeleton $X^{(n+1)}$ is obtained by attaching $(n+1)$-cells to $X^{(n)}$ by maps

$$\partial e^{n+1} \xrightarrow{f} X^{(n)}.$$

Denote by $\tilde{C}_n(X)$ the free abelian group generated by the oriented n-cells of X. Since each n-cell has two orientations, there are two generators in $\tilde{C}_n(X)$ for each n-cell. Introduce the relation that the two generators corresponding to the opposite orientations of a cell are negatives of one another. Denote the quotient group by $C_n(X)$. It is a free abelian group of rank equal to the number of n-cells.

If we are given an orientation for an n-cell e^n of X, then choosing an orientation preserving isomorphism $\Delta^n \to e^n$ determines an element in $H_n(X^{(n)}, X^{(n+1)})$. The resulting element depends only on the orientation of e^n and changes sign if we reverse the orientation. Thus, the construction determines a map $\varphi_n: C_n(X) \to H_n(X^{(n)}, X^{(n-1)})$. As we shall see below, an easy application of the excision theorem for homology shows that this mapping is an isomorphism. Let us assume for the moment that φ_n is an isomorphism. Define the boundary map $\partial: C_n(X) \to C_{n-1}(X)$ to be the composition:

$$C_n(X) \xrightarrow{\varphi_n} H_n(X^{(n)}, X^{(n-1)}) \xrightarrow{\partial} H_{n-1}(X^{(n-1)}) \xrightarrow{i_*} H_{n-1}(X^{(n-1)}, X^{(n-2)})$$

$$\xrightarrow{\varphi_{n-1}^{-1}} C_{n-1}(X).$$

Theorem 3.1 (CW Homology Theorem). $\{C_*(X), \partial\}$ *is a chain complex. Furthermore, there is a natural identification of its homology with the singular homology of X.*

P. Griffiths and J. Morgan, *Rational Homotopy Theory and Differential Forms*, Progress in Mathematics 16, DOI 10.1007/978-1-4614-8468-4_3,

© Springer Science+Business Media New York 2013

Remarks. (1) In case X has only finitely many cells in each dimension, the chain groups $C_n(X)$ are finitely generated, free abelian groups. Hence, the above theorem gives a fairly effective way of calculating homology.

(2) If (X, A) is a CW pair, i.e., if X is built by inductively attaching cells to A, then one has an analogously defined chain complex $\{C_n(X, A), \partial\}$ which computes the relative singular homology $H_*(X, A)$.

3.2 The Proof

Let us begin the proof of Theorem 2.1 with a couple of simple lemmas.

Lemma 3.2. *Let* $X = Y \cup_f e^m$. *Then*

$$\begin{cases} H_i(X, Y) = 0 & i \neq m, \\ H_m(X, Y) \cong \mathbb{Z}. \end{cases}$$

Proof. Let N be a neighborhood of ∂e^m in e^n and $U = Y \cup_f N$.

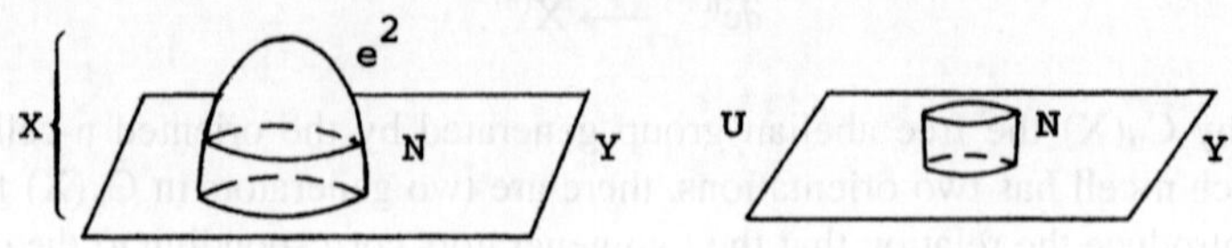

Then there is a deformation retraction of U onto Y (cf. the picture). The excision and homotopy axioms give

$$\begin{aligned} H_i(X, Y) &\cong H_i(X, \bar{U}) \cong H_i(\mathrm{int}(e^m), \mathrm{int}(e^n) \cap \partial N) \\ &\cong H_i(e^m, \partial e^m), \end{aligned}$$

where the first and third isomorphisms come from the fact that homotopy equivalent pairs have isomorphic homology and the second comes from excision. But $H_i(e^m)=0$ for $i > 0$, and so the exact homology sequence and computation of the homology of $S^{m-1} = \partial e^m$ give (for $i \neq 1$)

$$H_i(e^m, \partial e^m) \cong H_{i-1}(\partial e^m) = \begin{cases} \mathbb{Z} \text{ for } i = m \\ 0 \text{ otherwise.} \end{cases}$$

■

We define the reduced homology of a space X, denoted $\tilde{H}_*(X)$ to be the kernel of the map $H_*(X) \to H_*(pt)$ induced by the collapsing map $X \to pt$.

Remark. A similar proof gives the following useful result:

Proposition 3.3. *Let (X, A) be a CW pair. Let* X/A *denote the CW complex obtained by shrinking* A *to a point. Then* $H_i(X, A) \cong \tilde{H}_i(X/A)$.

The proof is by induction on the dimension of the cells and proceeds similarly to the proof of Lemma 2.2. Details are left to the reader.

It is excision which makes homology more computable than other homotopy functors such as the homotopy groups.

Lemma 3.4. *Let* X *be a CW complex. Then*

$$\begin{aligned} H_i(X, X^{(n-1)}) &= 0 \qquad && \text{for } i \leq n-1 \\ H_i(X^{(n)}, X^{(n-1)}) &= 0 \qquad && \text{for } i \neq n. \end{aligned}$$

The map $\varphi_n\colon C_n(X) \to H_n(X^{(n)}, X^{(n-1)})$ *is an isomorphism.*

Proof. These statements are all proved from Lemma 2.2 using induction on the number of cells (plus a direct limit argument if X has infinitely many cells). ■

Proof (Proof of Theorem 3.1). We have an isomorphism

$$\varphi_n\colon C_n(X) \longrightarrow H_n(X^{(n)}, X^{(n-1)})$$

and $\partial\colon C_n(X) \to C_{n-1}(X)$ is $\varphi_{n-1}^{-1} \circ \tilde{\partial} \circ \varphi_n$ where $\tilde{\partial}\colon H_n(X^{(n)}, X^{(n-1)}) \to H_{n-1}(X^{(n-1)}, X^{(n-2)})$ is the boundary in the homology long exact sequence for the triple $X^{(n-2)} \subset X^{(n-1)} \subset X^{(n)}$. Since $\tilde{\partial}_{n-1} \circ \tilde{\partial}_n = 0$, it follows that $\partial \cdot \partial\colon C_n(X) \to C_{n-2}(X)$ is zero. Thus, $\{C_*(X), \partial\}$ is a chain complex. Setting $Z = X^{(n-2)}$ and applying Lemma 2.2, we have an exact sequence

$$\begin{array}{ccccccc} 0 & \longrightarrow & H_n(X^{(n)}, Z) & \longrightarrow & H_n(X^{(n)}, X^{(n-1)}) & \xrightarrow[\tilde{\partial}]{} & H_{(n-1)}(X^{(n-1)}, Z) \\ & & \downarrow{=} & & \downarrow{=} & & \downarrow{=} \\ 0 & \longrightarrow & H_n(X^{(n)}) & \longrightarrow & C_n(X) & \xrightarrow[\partial]{} & C_{n-1}(X). \end{array}$$

Thus, $H_n(X^{(n)})$ is identified with the cycles in $C_n(X)$. Similarly, there is a commutative diagram:

$$\begin{array}{ccccc} C_{n+1}(X) & \xrightarrow{=} & H_{n+1}(X^{(n+1)}, X^{(n)}) & \xrightarrow{\partial} & H_n(X^{(n)}) \\ & & \downarrow & \nearrow_{\partial} & \\ & & H_{n+1}(X, X^{(n)}) & & \\ & & \downarrow & & \\ & & 0 & & \end{array}$$

This allows us to identify $\partial C_{n+1}(X) \subset Z_n(C_*(X))$ with $\partial(H_{n+1}(X, X^{(n)})) \subset H_n(X^{(n)})$. Applying 2.2 and 2.4 once again, we find the following is exact:

$$H_{n+1}(X, X^{(n)}) \xrightarrow{\partial} H_n(X^{(n)}) \longrightarrow H_n(X) \longrightarrow 0.$$

Thus, $Z_n(C_*(X))/\partial C_{n+1}(X) = H_n(X^{(n)})/\partial(H_{n+1}(X, X^{(n)})) = H_n(X)$. If $f: X \to Y$ is a continuous mapping between CW complexes, then f is homotopic to a cellular map. If we deform f to be cellular, then it induces maps on all the relative groups used in establishing the equivalent of $H_*(C_*(X))$ with $H_*(X)$. Hence, this equivalence is functorial. ■

Remark. Choose an orientation for each n-cell and each (n + 1)-cell of X. If e_α^{n+1} is an (n + 1)-cell, then $\partial e_\alpha^{n+1} = \Sigma_\beta c_{\alpha\beta} e_\beta^n$ for some integer coefficients $c_{\alpha\beta}$. This means that if $f_\alpha: \partial e_\alpha^{n+1} \to X^{(n)}$ is the attaching map for e_α^{n+1}, then there is a singular chain c in $X^{(n-1)}$ so that $\partial e_\alpha^{n+1} - \Sigma c_{\alpha\beta} e_\beta^n - c$ is a singular boundary in $X^{(n)}$.

If the attaching maps are generic, then there is an appealing geometric description of $\partial: C_n(X) \to C_{n-1}(X)$. Choose an oriented n-cell $(e^n, \mathcal{O}_e)$ and an oriented (n − 1)-cell $(\sigma^{n-1}, \mathcal{O}_\sigma)$. To calculate the coefficient of $(\sigma^{n-1}, \mathcal{O}_\sigma)$ in $\partial(e^n, \mathcal{O}_e)$, one simply considers the attaching map $\partial e^n \xrightarrow{\varphi_e} X^{(n-1)}$. Generically this map will be transverse to some point p_σ in σ^{n-1}. Comparison of the orientations allows us to assign an algebraic intersection number ±1 to each point in $\varphi_e^{-1}(p_\sigma)$. The sum of these is the sought-after coefficient.

3.3 Examples

1. $\mathbb{R}P^2 = S^1 \cup_{z^2} D^2$ where $D^2 = \{z \in \mathbb{C}: |z| \le 1\}$.
   The boundary map sends $[D^2, \mathcal{O}_{D^2}] \to 2[e^1, \mathcal{O}_{e^1}]$.
2. The 2-torus is obtained by attaching D^2 to $S^1 \vee S^1$ by $aba^{-1}b^{-1}$

   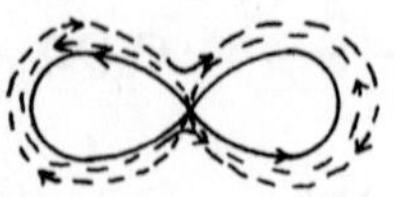

   $\varphi^{-1}(P_a) = 2$ points with opposite sign
   $\varphi^{-1}(P_b) = 2$ points with opposite sign.
   Hence, $\partial[D^2, \mathcal{O}_{D^2}] = 0$.
3. The n-sphere $S^n = x_0 \cup_f e^n$. Thus, $H_i(S^n) = 0$ for $i \ne 0, n$ and $H_n(S^n) = \mathbb{Z}$.
4. Complex projective space $\mathbb{C}P^n$ is obtained from $\mathbb{C}P^{n-1}$ by attaching a cell of dimension 2n. Inductively, one argues that $\mathbb{C}P^n$ has a cell decomposition with exactly one cell occurring in dimensions $0, 2, 4, \ldots, 2n$. Thus, in $C_*(X)$, all boundary maps must be zero. Hence,

$$H_{2i}(\mathbb{C}P^n) = \mathbb{Z} \text{ for } 0 \le i \le n$$

$$H_*(\mathbb{C}P^n) = 0 \quad \text{for all other } * .$$

5. Consider the map $f_n: S^1 \to S^1$ given by $e^{i\theta} \to e^{in\theta}$ for $n \in \mathbb{Z}$. Let $X_n = S^1 \cup_{f_n} e^2$. Then

$$H_1(X_n) = \mathbb{Z}/n\,\mathbb{Z}.$$

The reason is that $\partial e^2 = nS^1$ (cf. example 1 for $n = 2$).
6. As we saw in Chap. 2, a simplicial complex has a natural CW complex structure. The CW homology theorem gives a purely combinatorial way to calculate $H_*(|K|)$. (Here $|K|$ is the topological space associated to K.) The complex $\{C_*(|K|), \partial\}$ is exactly the simplicial chain complex.
7. If (X, A) is a CW pair, then there is an analogously defined complex $\{C_n(X, A), \partial\}$. The group $C_n(X, A)$ is a free abelian group with one generator for each n-cell in X–A. Arguments similar to the ones above show that the homology of this complex is naturally identified with $H_*(X, A)$.

Chapter 4
The Whitehead Theorem and the Hurewicz Theorem

4.1 Definitions and Elementary Properties of Homotopy Groups

For spaces X and Y, we denote by [X, Y] the set of homotopy classes of maps from X to Y. For pairs of topological spaces (X, A) and (Y, B), we denote by [(X, A), (Y, B)] the set of homotopy classes of maps of the pair (X, A) to the pair (Y, B). If H is such a homotopy, then for each t, the map H_t is required to map A to B. If (X, x_0) and (Y, y_0) are based spaces, then we denote $[(X, x_0), (Y, y_0)]$ by $[X, Y]_0$. There is a similar notation homotopy classes of maps of triples $[(X, A, A'), (Y, B, B')]$ where $A' \subset A \subset X$ and $B' \subset B \subset Y$

Recall that the set $[S^1, X]_0$ has a group structure induced by composing loops. This group is the *fundamental group* and is denoted $\pi_1(X, x_0)$ (or $\pi_1(X)$ if the base point plays no important role). We define sets $\pi_n(X, x_0) = [S^n, X]_0$. Then $\pi_0(X)$ is exactly the set of path components of X. We show how to make $\pi_n(X)$ an abelian group for $n \geq 2$.

To define the group structure on $\pi_n(X), n \geq 2$ and show that it is abelian, it is convenient to give a different description. Let I^n denote the n-cube. Then $\pi_n(X) = [\,(I^n, \partial I^n); (X, x_0)]$. The composition is

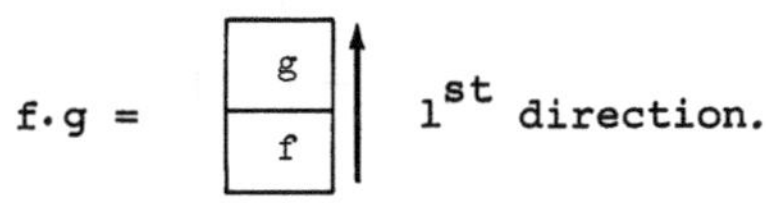

This is abelian as the following sequence of pictures indicates (the unmarked areas are sent to the base point).

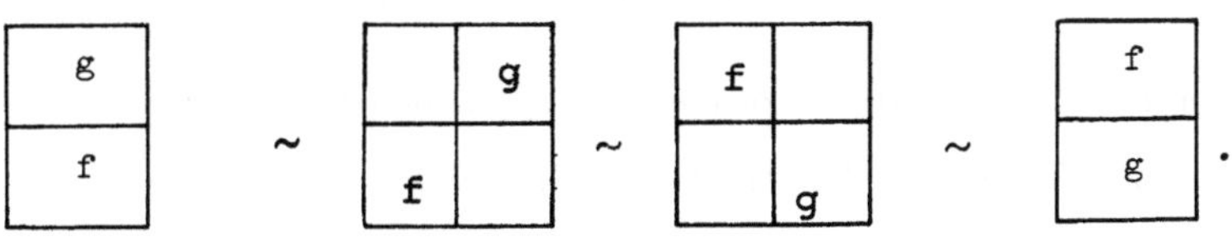

P. Griffiths and J. Morgan, *Rational Homotopy Theory and Differential Forms*, Progress in Mathematics 16, DOI 10.1007/978-1-4614-8468-4_4,

© Springer Science+Business Media New York 2013

For $n \geq 2$, define $\pi_n(X, A)$ to be

$$[(I^n, \partial I^n, \overline{\partial I^n - I^{n-1} \times \{1\}}), (X, A, a_0)].$$

The group law is given as before. An argument similar to the one above shows that for $n \geq 3$, $\pi_n(X, A)$ is an abelian group. Given $f \in [(X, A), (Y, B)]$, there is an induced map

$$f_*\colon \pi_*(X, A) \to \pi_*(Y, B).$$

This means that the homotopy groups form a covariant functor from the homotopy category (of pairs) of spaces to the category of groups.

Remarks. (i) The homotopy groups are easier to define but much harder to calculate than the homology groups. One of the main points of this course will be to give an effective way to calculate $\pi_* \otimes \mathbb{Q}$ from the DGA of differential forms.

(ii) We show that $\pi_k(S^n) = 0$ if $k < n$ and $\pi_3(S^2) \neq 0$. To see the first, let $f\colon S^k \to S^n$ be given with $k < n$. Deform f until it is cellular. Since the k-skeleton of S^n can be taken to be a point, this deformation carries f to a constant map. (In general, this argument shows that $\pi_k(X^{(m)}) \xrightarrow{\sim} \pi_k(X)$ for $k \leq m - 1$ and that $\pi_k(X^{(k)}) \to \pi_k(X)$ is onto.) To show that $\pi_3(S^2) \neq 0$, consider $\mathbb{C}P^2$ as an adjunction space $\mathbb{C}P^2 = (\mathbb{C}P^1) \cup_f e^4$ where $\mathbb{C}P^1 = S^2$. The homotopy type of an adjunction space depends only on the homotopy class of the attaching map. Thus, if $\pi_3(S^2) = 0$, then $\mathbb{C}P^2$ would be homotopy equivalent to $S^2 \cup_{\text{constant}} e^4 = S^2 \vee S^4$. This would imply that if g generates $H^2(\mathbb{C}P^2)$, then $g \cup g = 0$ in $H^4(\mathbb{C}P^2)$. But we know (say by Poincaré duality) that $g \cup g \neq 0$.

There is an analogous argument showing that $\pi_{4n-1}(S^{2n}) \neq 0$.

If $f\colon S^n \to X$, then there is induced $f_*\colon H_n(S^n) \to H_n(X)$. But $H_n(S^n)$ is identified, once and for all, with $\mathbb{Z}$. Thus, we have $f_*(1) \in H_n(X)$. This determines a natural transformation:

$$H\colon \pi_n(X) \longrightarrow H_n(X)$$
$$[f] \longrightarrow f_*(1)$$

called the *Hurewicz homomorphism.* (It is an easy exercise to show that H is indeed a homomorphism.) Applying this in the special case where $X = S^n$, we have

$$H\colon \pi_n(S^n) \longrightarrow H_n(S^n) = \mathbb{Z}.$$

H(f) is also called the *degree of* f and sometimes denoted by deg(f).

Theorem 4.1 (Brouwer). *The map* $H\colon \pi_n(S^n) \to \mathbb{Z}$ *is an isomorphism.*

Since $H(Id_{S^n}) = 1$, it is obvious that H is onto. It is harder to show that H is one-to-one, i.e., that if $\deg(f) = \deg(g)$, then $f \simeq g$. We shall prove this later in this chapter.

If $x_0 \in A \subset X$, then there is an exact homotopy sequence

$$\ldots \longrightarrow \pi_{n+1}(X, A) \xrightarrow{\partial} \pi_n(A) \xrightarrow{i_*} \pi_n(X) \xrightarrow{j_*} \pi_n(X, A) \longrightarrow \ldots$$

As one consequence of this and the calculations above, we see that excision is false for the homotopy groups.

The easiest example of this is that $\pi_3(D^2, S^1) \overset{\partial}{\underset{\sim}{\rightarrow}} \pi_2(S^1)$ is zero whereas $\pi_3(S^2,\ D^2) \cong \pi_3(S^2) \neq 0$. Of course, where excision to hold for homotopy groups, then they would satisfy all the axioms for homology. If that were true, then the argument in Chap. 3 which identified singular homology with $H_*(C_*(X, A))$ could be used to identify the homotopy groups with $H_*(C_*(X, A))$. It would follow that the homology and homotopy groups were isomorphic.

4.2 The Whitehead Theorem

Theorem 4.2. *Let X and Y be CW complexes with base points,* x_0 *and* y_0*, being 0-cells. Let* $f: (X, x_0) \to (Y, y_0)$ *be a map inducing isomorphisms*

$$f_*: \pi_n(X, x_0) \xrightarrow{\cong} \pi_n(Y, y_0) \quad \textit{for all } n \geq 0$$

Suppose that Y is connected. Then $f: X \to Y$ *is a homotopy equivalence.*

Let us begin with a special case. Suppose $\dim X < \infty$ and $\pi_n(X) = 0$ for all $n \geq 0$. We shall show that X is contractible. (This is a special case of the theorem for $x_0 \hookrightarrow X$.) Since $\pi_0(X)$ is the one point set, the zero skeleton can be deformed to $x_0 \in X$.

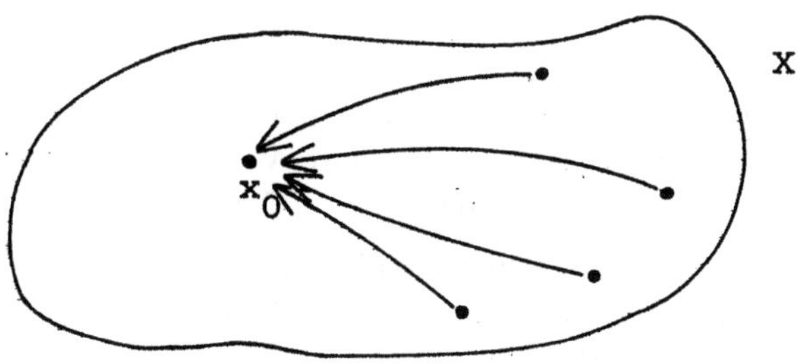

Use the homotopy extension property (Theorem 2.1) to obtain a continuous family:

$$f_t: X \longrightarrow X \quad 0 \leq t \leq 1$$

such that $f_0 = Id_X$ and $f_1(X^{(0)}) = x_0$.

Now consider $X^{(1)}$. The image under f_1 of any 1-cell e^1 is a loop based at x_0. Since $\pi_1(X) = 0$ we can deform $f_1(e^1)$ through loops based at x_0 to the constant loop. Doing this for each 1-cell defines a homotopy from $f_1|X^{(1)}$ to the constant map $f_2: X^{(1)} \to x_0$. It is a homotopy relative to $X^{(0)}$. Using homotopy extension, we can find a homotopy $f_t: X \to X, 1 \leq t \leq 2$, with $f_2|X^{(1)} =$ constant. Continue in this fashion using the fact that $\pi_n(X) = 0$ for all n. In the end, we have a homotopy from Id_X to the constant map of X to $x_0 \in X$.

Remarks. (i) The same argument shows that if $\pi_n(X) = 0$ for $n < N$, then there is a homotopy

$$f_t: X \longrightarrow X; \qquad 0 \leq t \leq 1$$

with

$$f_0 = Id$$

and

$$f_1(X^{(N-1)}) = x_0.$$

As a consequence, $\tilde{H}_k(X) = 0$ for $k < N$.

(ii) In case X is infinite dimensional, we make the first homotopy last for $0 \leq t \leq 1/2$, the second last for $1/2 \leq t \leq 3/4$, etc. and then define $f_1(X) = x_0$. Using the fact that the topology on X is the weak (or limit) topology, it is an easy exercise to show that the proposed homotopy is indeed continuous.

(iii) There is a relative version of the theorem which states that if (X, A) is a CW pair and if $\pi_n(X, A) = 0$ for all n, then there is a deformation retraction

$$f_1: X \longrightarrow A,$$

i.e., there is a homotopy $f_t: X \to X$; $0 \leq t \leq 1$, such that $f_0 =$ identity, $f_1(X) \subset A$, and $f_t|A = \text{identity}_A$. The proof is essentially the same as in the absolute case.

General Case: Given $f: X \to Y$, choose a cellular map f' homotopic to f. The pair $(M_{f'}, Y)$ is a relative CW complex. Since $M_{f'}$ retracts to Y, it follows that $\pi_k(M'_f) = \pi_k(Y)$ for all k. If $f'_*: \pi_k(X) \to \pi_k(Y)$ is an isomorphism for all k, then so is $\pi_k(X) \xrightarrow{i_*} \pi_k(M_{f'})$. Hence, $\pi_k(M_{f'}, X) = 0$ for all k. Thus, remark (iii) implies that there is a retraction $r: M_{f'} \geq X$. The composition $Y \subset M_{f'} \xrightarrow{r} X$ is a homotopy inverse for f' and hence for f. This completes the proof of the Whitehead theorem.

Example. Think of S^k as $I^k/\{\partial I^k\}$. There is a product map

$$I^2/\{\partial I^2\} \times I^2/\{\partial I^2\} \longrightarrow I^4/\{\partial I^4\}.$$

This is a map $f: S^2 \times S^2 \to S^4$. Since $\pi_k(S^2 \times S^2) \cong \pi_k(S^2) \times \pi_k(S^2)$, one sees easily that $f_*: \pi_k(S^2 \times S^2) \to \pi_k(S^4)$ is trivial for all k. But f is <u>not</u> homotopic to a constant map. The easiest way to see this is to note that $f_*: H_4(S^2 \times S^2) \to H_4(S^4)$ is nonzero (in fact, it is an isomorphism).

In dealing with spaces that are not CW complexes, a map $f: X \to Y$ is not necessarily a homotopy equivalence if it induces an isomorphism in all homotopy groups. When it does, we say that f is a *weak homotopy equivalence*. This is not an equivalence relation, but it generates an equivalence relation weak homotopy equivalence which when restricted to CW complexes is homotopy equivalence by Whitehead's theorem. In general if A is a CW complex and $f: X \to Y$ is a weak homotopy equivalence, then $f_\#: [A, X] \to [A, Y]$ is a bijection.

4.3 Completion of the Computation of $\pi_n(S^n)$

Theorem 4.3. *The Hurewicz homomorphism $\pi_n(S^n) \to H_n(S^n)$ is an isomorphism.*

Proof. Since we have seen that $H: \pi_n(S^n) \xrightarrow{\sim} \mathbb{Z}$ is onto, it suffices to show that if $f_*(1) = 0$ in $H_n(S^n)$, then $f \sim$ constant. First, deform f until it is a C^∞ mapping and $p \in S^n$ is a regular value for f. This means that f is a local diffeomorphism in a neighborhood of each $x_i \in f^{-1}(p)$.

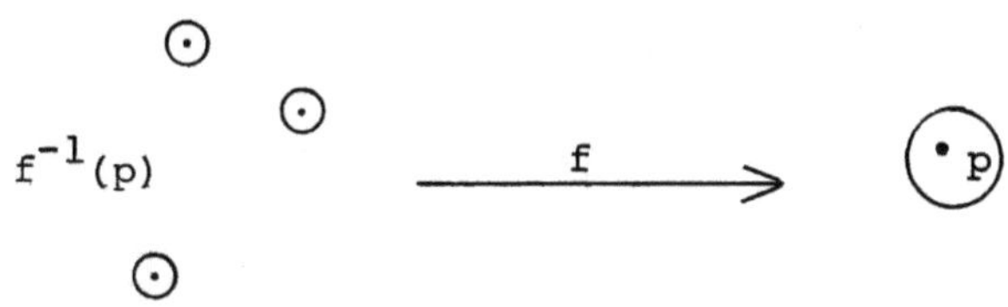

At each point $x_i \in f^{-1}(p)$, there is a local degree of f. This is $+1$ if f is orientation-preserving near x_i and -1 if it is orientation reversing there.

We claim that

$$\deg(f) = \Sigma_{x_i \in f^{-1}(p)} \epsilon_f(x_i)$$

(where $\epsilon_f(x_i)$ is the local degree of f at x_i). The easiest way to see this is to choose a C^∞-form ω on s^n of degree n supported in a small ball $D^n \subset S^n$ containing p whose integral over D^n is 1. Clearly, $\int_{S^n} f^*(\omega) = \deg(f)$. On the other hand, if we choose D^n sufficiently Small, there will be one component of $f^{-1}(D^n)$ for each $x_i \in f^{-1}(p)$. If C_i is the component of $f^{-1}(D)$ containing x_i, then

$$\int_{C_i} f^*(\omega) = (\deg(f) \text{ at } x_i) \cdot \int_{Dn^\omega} = \deg(f) \text{ at } x_i.$$

The result follows immediately.

This argument assumes deRham's theorem from Chap. 9. Alternatively one may use

$$H_n(S^n) \cong H_n(S^n, p) \cong H_n(S^n, D^n)$$
$$\cong H_n(S^n - D^n, \partial D^n) \cong H_{n-1}(S^{n-1})$$

together with

$$H_n(S^n) \cong H_n(S^n, f^{-1}(p)) \cong H_n(S^n, f^{-1}(D^n))$$
$$\cong H_n(S^n - f^{-1}(D^n), f^{-1}(\partial D^n)) \hookrightarrow \oplus_i H_{n-1}(\partial C_i)$$

to show that

$$\deg f = \Sigma_i \deg(f|\partial C_i).$$

Thus, we must show that if $\Sigma_{x_i \in f^{-1}(p)}$ (deg (f) at x_i) $= 0$, then f is homotopic to constant. We do this for $n \geq 2$ (the case $n = 1$ will be proved afterwards). Assume first that $f^{-1}(p) = \{x\} \cup \{y\}$, that deg(f) $= 1$ at x, and that deg(f) $= -1$ at y. In $S^n \times I$, choose an embedded arc A connecting x and y and meeting $S^n - \{0\}$ transversally at x and y

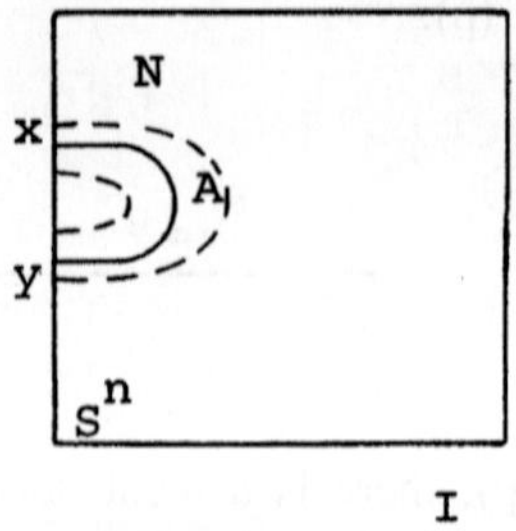

The disks around x and y [i.e. $f^{-1}(D^n)$] extend to a tubular neighborhood N of A. In fact, $N \cong e^n \times I$

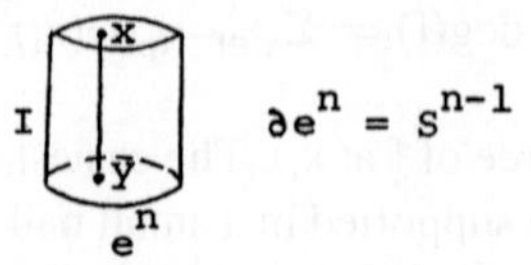

Note that f is defined on $e^n \times \{0\}$ and $e^n \times \{1\}$ and, since the local degree of f at x and y has opposite signs, the degrees of f on $S^{n-1} \times \{0\}$ and $S^{n-1} \times \{1\}$ are the same. By the induction assumption, f extends to a map $F: S^{n-1} \times I \to \partial D^n$. We may extend F to a map $F: e^n \times I \to D^n$ (this is a little exercise). Thus, f extends to a map $\tilde{f}: S^n \times \{0\} \cup N \to S^n$ such that $\tilde{f}: \partial N \to \partial D^n \subset S^n$. Since $S^n \setminus \operatorname{int} D^n$ is contractible, we can extend $\tilde{f}$ to a map on the rest of $S^n \times I$, sending the complement of N to D^n. This map will have p as a regular value with preimage the arc A. The resulting map on $S^n \times \{1\}$ will miss p and hence is homotopic to a constant.

If $f^{-1}(p)$ has more than two points and $n \geq 2$, then this argument allows us to deform f by a homotopy to "cancel" two of the points with opposite local degree. Continuing in this manner, we can deform f until the preimage of p is empty if the deg(f) is 0. Once we have accomplished this, the result is clearly homotopic to a constant map.

If $n = 1$, then we must take more care. The point is that the arc A between two points of opposite sign must be disjoint from $x_j \times I$ for all other points $x_j \in f^{-1}(p)$. If $n \geq 2$, then general position ensures that this is always possible. If $n = 1$, then we must find a pair of points x and $y \in f^{-1}(p)$ of opposite local degree so that there are no other points of $f^{-1}(p)$ "in between them" (i.e., so that one of the two arcs in S^1 connecting them contains no other points of $f^{-1}(p)$). Since this is always possible if $\deg(f) = 0$, we can still cancel a pair of points in this case. ∎

There is a generalization of this result that we shall need.

Theorem 4.4. *For* $n > 1$, *the map* $H\colon \pi_n(\vee_i S^n) \to H_n(\vee_i S^n) \cong \oplus_i \mathbb{Z}$ *is an isomorphism.*

Proof. The proof is analogous to the one given above. One takes $f\colon S^n \to \vee_i S^n_i$ and deforms it until it is regular with respect to a point $p_i \in S^n_i$. (The point p_i is different from the point at which all the S^n_i are joined together.) The cancelling argument shows that if the sum of the local degrees at $x_{ij} \in f^{-1}(p_i)$ is zero, then f is homotopic to a map missing p_i. (This requires that $n \geq 2$.) If $H(f) = 0$, then all the local degrees $\Sigma_j(\deg(f) \text{ at } x_{ij}) = 0$ for all i. Thus, if $H(f) = 0$, then f is homotopic to a map missing a point $p_i \in S_i$ for every i. This map is in turn homotopic to the constant map since $\vee_i(S^i - p_i)$ is contractible. ∎

4.4 The Hurewicz Theorem

We begin with the statement of this result.

Theorem 4.5. *Let X be a CW complex. If* $\pi_k(X) = 0$ *for* $k < n$, *then*

(i) $\tilde{H}_k(X) = 0$ *for* $k < n$ *(recall that* $\tilde{H}$ *is the reduced homology), and*
(ii) $H\colon \pi_n(X) \to H_n(X)$ *is an isomorphism provided* $n > 1$.

Proof. We have already seen that if $\pi_k(X) = 0$ for $k < n$, then $\tilde{H}_k(X) = 0$ for $k < n$. It remains to show that under this hypothesis, we have

$$H\colon \pi_n(X) \xrightarrow{\sim} H_n(X).$$

Step I: Let us show that H is onto. Since $\pi_i(X) = 0$ for $i < n$, the argument given in the proof of the Whitehead theorem shows that there is a map $f\colon X \to X$ such that (1) f is homotopic to the identity and (2) $f(X^{(n-1)}) = x_0$. Let $a \in H_n(X)$, and let $\Sigma_{n-\text{cells}} a_\alpha(e^n_\alpha, \mathcal{O}_\alpha)$ be a cycle representative for a in $C_n(X)$. (Here, we choose arbitrarily an orientation $\mathcal{O}_\alpha$ for each e^n_α). Clearly $f|e^n_\alpha\colon (e^n_\alpha, \partial e^n_\alpha) \to (X, x_0)$. Thus,

$f|e^n_\alpha$ represents an element in $\pi_n(X, x_0)$. Since f is homotopic to the identity, $\alpha = f_*(\alpha) =$ class of $\Sigma_\alpha f(e^n_\alpha)$. Since each $f(e^n_\alpha)$ is represented by a sphere, this latter class is in the image of H.

We wish to show that $H: \pi_n(X) \to H_n(X)$ is injective, provided that $n > 1$ and $\pi_i(X) = 0$ for $i < n$. Let $f: S^n \to X$ be a map such that $f_*[S^n] = 0$ in $H_n(X)$. We can deform f until $f(S^n) \subset X^{(n)}$ and until f is transverse regular to a point p_i in the interior of each n-cell e^n_i. Let

$$\lambda_i = \Sigma_{x \in f^{-1}(p_i)} \text{ (local deg at } x).$$

The chain $\Sigma\, \lambda_i[e^n_i]$ is a cycle in $C_n(X)$ which represents $f_*[S^n] \in H_n(X)$. Since $f_*[S^n] = 0$, this cycle is a boundary; i.e., there exists μ_j such that $\Sigma\lambda_i[e^n_i] = \partial\Sigma\mu_j e^{n+1}_j$. Adding to $f: S^n \to X^n$ the linear combination $\Sigma\mu_j\partial e^{n+1}_j$ makes each $\lambda_i = 0$, and this does not change the homotopy class of $f: S^n \to X$. There is a map $\psi: X \to X$ such that (a) ψ is homotopic to id_X and (b) $\psi|X^{(n-1)}$ is constant. This means that $\psi|X^{(n)}$ factors through a wedge of spheres:

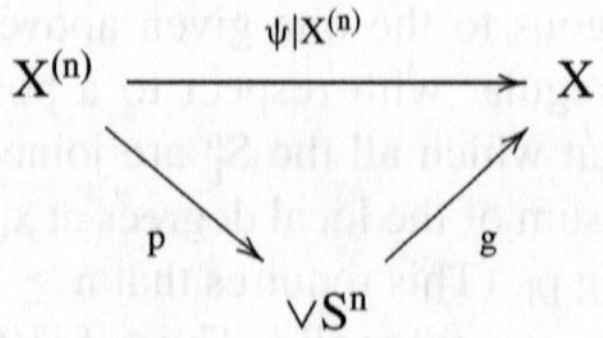

Since $f: S^n \to X^{(n)}$ has each $\lambda_i = 0$, the composition $p \circ f: S^n \to \vee S^n$ is homologous to zero. By Lemma 3.2, this means that $p \circ f: S^n \to \vee S^n$ is homotopic to zero. Hence, $g \circ p \circ f = \psi \circ f$ is homotopically trivial. Since $\psi \cong id_X$, this means that f itself is homotopically trivial. ■

Note: The Hurewicz theorem is valid for all spaces and not just CW complexes. To establish this, one works with the cycles themselves rather than with the spaces into which they are mapped. For a proof of the general result, see [23].

4.5 Corollaries of the Hurewicz Theorem

To begin with, there is (of course) a relative form.

Let $A \subset X$ be a CW pair and assume that

$$\pi_1(A) = 0$$

and

$$\pi_i(X, A) = 0, \quad i < n \ (n \geq 2).$$

Then it follows that

$$H_i(X, A) = 0 \quad \text{for } i < n$$

and the relative Hurewicz map

$$\pi_n(X, A) \xrightarrow[H]{} H_n(X, A)$$

is an isomorphism (the proof that $H_i(X, A) = 0$ for $i < n$ and that H is surjective follows from the proof of the Whitehead theorem, just as in the absolute case).

We now come to the corollaries of Hurewicz theorem.

Corollary 4.6. *If* X *and* Y *are simply connected CW complexes and* $X \xrightarrow{f} Y$ *induces an isomorphism on homology, then* f *is a homotopy equivalence.*

Proof. Let M_f be the mapping cylinder for f. Then $Y \hookrightarrow M_f$ is a deformation retract, and using this, we make the identifications

$$\pi_i(Y) \cong \pi_i(M_f)$$
$$H_i(Y) \cong H_i(M_f).$$

The inclusion $X \hookrightarrow M_f$ gives exact sequences

$$\begin{array}{ccccccccc}
\cdots \longrightarrow & \pi_i(X) & \xrightarrow{f_*} & \pi_i(Y) & \longrightarrow & \pi_i(M_f, X) & \longrightarrow & \pi_{i-1}(X)\cdots \\
& \downarrow & & \downarrow & & \downarrow & & \downarrow \\
\cdots \longrightarrow & H_i(X) & \xrightarrow{f_*} & H_i(Y) & \longrightarrow & H_i(M_f, X) & \longrightarrow & H_{i-1}(X)\cdots
\end{array}$$

Using that f_* is an isomorphism on homology gives $H_i(M_f, X) = 0$ for $i \geq 0$. The relative Hurewicz theorem now gives

$$\pi_i(M_f, X) = 0 \quad \text{for } i \geq 0.$$

It follows then that M_f deformation retracts onto X (cf. the proof of the Whitehead theorem). ■

Corollary 4.7. *If* X *has the homotopy type of an* n*-dimensional CW complex and if* $\pi_i(X) = 0$ *for* $i \leq n$*, then* X *is contractible.*

Proof. Since $\pi_i(X) = 0$ for $i \leq n$, $H_i(X) = 0$ for $i \leq n$. On the other hand, $H_i(X) = 0$ for $i > n$ by dimension. Thus, $\tilde{H}_*(X) = 0$ and so $\pi_*(X) = 0$ by the Hurewicz theorem. Hence, by the Whitehead theorem, X is contractible. ■

Corollary 4.8. *If* X *has the homotopy type of an* n*-dimensional CW complex and if* $\pi_i(X) = 0$ *for* $i \leq n - 1$*, then*

$$X \sim \vee S^n.$$

In particular, if $H_n(X) \cong \mathbb{Z}$*, then* X *has the homotopy type of* S^n*. (Thus, a simply connected homology sphere (a homology sphere is a space with the same homology as a sphere) is homotopy equivalent to a sphere.)*

Proof. This follows from the proof of the Hurewicz theorem in the following way: We may assume that $X = X^{(n)}$. As before, there is a map $\psi: X \to X$ homotopic to the identity with $\psi(X^{(n-1)})$. Thus, ψ factors

$$X \xrightarrow{g} \vee S^n \xrightarrow{f} X.$$

It follows that $H_n(\vee S^n) \xrightarrow{f_*} H_n(X)$ is onto and we may choose a basis for $H_n(\vee S^n) \cong \oplus H_n(S^n)$ (little exercise) and $H_n(X)$ such that f_* is given by a matrix with c columns and b rows, where b is the rank of $H_n(X)$, and c is the rank of $H_n(\vee S^n)$, and each a_i is ± 1:

$$\begin{pmatrix} a_1 & 0 & \cdot & \cdot & \cdot & 0 \cdots 0 \\ 0 & a_2 & 0 & \cdot & \cdot & 0 \cdots 0 \\ 0 & 0 & \cdot & 0 & 0 & 0 \cdots 0 \\ 0 & \cdot & 0 & \cdot & 0 & 0 \cdots 0 \\ 0 & \cdot & \cdot & 0 & a_b & 0 \cdots 0 \end{pmatrix}.$$

Since $\pi_n(\vee S^n) \cong H_n(\vee S^n)$, we may find a map h

$$\begin{array}{ccc} S^n \vee \cdots \vee S^n & & \\ {\scriptstyle h}\downarrow & \searrow^{j} & \\ \vee S^n & \xrightarrow[f]{} & X \end{array}$$

such that h_* has matrix with b columns and c rows:

$$\begin{pmatrix} a_1 & 0 & \cdot & \cdot & \cdot \\ 0 & a_2 & 0 & \cdot & \cdot \\ 0 & 0 & \cdot & 0 & 0 \\ 0 & \cdot & 0 & \cdot & 0 \\ 0 & \cdot & \cdot & 0 & a_b \\ 0 & 0 & \cdot & \cdot & 0 \end{pmatrix}$$

using the obvious basis for $H_n(\underbrace{S^n \vee \ldots \vee S^n}_{b})$.

It follows that j_* is an isomorphism on homology. ■

Corollary 4.9. *If* X *is simply connected and* $H_i(X) = 0$ *for* $i > n$*, then* $X \sim Y^{(n+1)}$*; i.e.,* X *has the homotopy type of an* $(n+1)$*-dimensional CW complex. If, in addition,* $H_n(X)$ *is free, then* $X \sim Y^{(n)}$.

Proof. We have from the CW homology theorem

$$H_i(X^{(n)}) \xrightarrow{\cong} H_i(X) \qquad i \neq n$$
$$H_i(X^{(n)}) \longrightarrow H_n(X) \to 0$$

Using the exact homology sequence

$$H_i(X^{(n)}) \longrightarrow H_i(X) \longrightarrow H_i(X, X^{(n)}) \longrightarrow H_{i-1}(X^{(n)}) \longrightarrow H_{i-1}(X)$$

it follows that

$$H_i(X, X^{(n)}) = 0 \quad \text{for } i \neq n+1$$

and that

$$0 \longrightarrow H_{n+1}(X, X^{(n)}) \longrightarrow H_n(X^{(n)}) \xrightarrow{i*} H_n(X) \longrightarrow 0$$

is exact. Since $H_n(X^{(n)})$ is the kernel of the map from $C_n(X^{(n)} \to C_{n-1}(X^{(n)}$, it follows that $H_n(X^{(n)})$ is free abelian.

Using the relative Hurewicz theorem, we see that

$$\pi_{n+1}(X, X^{(n)}) \cong H_{n+1}(X, X^{(n)}).$$

Thus, we may attach $(n+1)$-cells to $X^{(n)}$ to in such a way that the attaching maps form a basis for the kernel of i_*. For this new CW complex $Y^{(n+1)}$,

$$H_i(Y^{(n+1)}) \cong H_i(X)$$

by construction for all i. Hence, $Y^{(n+1)})$ and X are homotopy equivalent. The other statement is an exercise. ■

Corollary 4.10. *Let X be a simply connected topological* n*-manifold. Then* X *has the homotopy type of an* n*-dimensional CW complex.*

Remark. It is pretty clear that X has the homotopy type of a simplicial complex [cf. Exercise (13)]. Thus, in the homotopy category, X can be triangulated. If X is smooth, then X can be smoothly triangulated (see pp. 124–125 of [27]). If X is a topological manifold, then there is exactly one obstruction to it being triangulated in a homogeneous fashion, an obstruction in $H^4(X; \mathbb{Z}/2\mathbb{Z})$; see [12]. (Here, "homogeneous fashion" means that every point has a neighborhood whose closure is a subcomplex that is combinatorially isomorphic to a subcomplex of a rectilinear triangulation of Euclidean space of the same dimension.) There are topological manifolds admitting no triangulation, homogeneous or not; see [15,22].

4.6 Homotopy Theory of a Fibration

Let $\pi: E \to B$ be a fibration (i.e., π has the homotopy lifting property). Let $\gamma: I \to B$ be a path from b_0 to b_1. We have a commutative diagram

$$\begin{array}{ccccc} \pi^{-1}(b_0) \times \{0\} & \hookrightarrow & & & E \\ \downarrow & & & & \downarrow \\ \pi^{-1}(b_0) \times I & \xrightarrow{p_2} & I & \xrightarrow{\gamma} & B \end{array}$$

Applying homotopy lifting gives a map $\pi^{-1}(b_0) \times I \to E$ covering $\gamma \circ p_2$. It is easy to show that $\pi^{-1}(b_0) \times \{1\} \to \pi^{-1}(b_1)$ is a homotopy equivalence. Thus, we see that if B is path connected, then all fibers are homotopy equivalent.

If γ is a loop based at b_0, then the resulting homotopy equivalence $\pi^{-1}(b_0) \times \{1\} \to \pi^{-1}(b_0)$ is a homotopy automorphism of $\pi^{-1}(b_0)$. Its homotopy class depends only on the class of γ in $\pi_1(B, b_0)$. Thus, we have a representation $\pi_1(B, b_0) \to \mathrm{Auto}(\pi^{-1}(b_0))$ where Auto $(\pi^{-1}(b_0))$ is the group of homotopy classes of homotopy equivalences of $\pi^{-1}(b_0)$. This representation is the *action of* $\pi_1(B, b_0)$ *as homotopy classes of homotopy equivalences on the fiber*. There are induced actions on the homology and cohomology of the fiber.

Theorem 4.11. *Let* $\pi: E \to B$ *be a fibration, and let* $F = \pi^{-1}(b_0)$. *There is an exact sequence:*

$$\longrightarrow \pi_{n+1}(B, b_0) \xrightarrow{\partial} \pi_n(F, e_0) \xrightarrow{i_*} \pi_n(E, e_0) \xrightarrow{\pi_*} \pi_n(B, b_0) \longrightarrow$$

where $i: F \hookrightarrow E$ *is the inclusion.*

Proof. We have the long exact sequence of the pair $F \hookrightarrow E$.

$$\cdots \longrightarrow \pi_{n+1}(E, F) \xrightarrow{\partial} \pi_n(F) \longrightarrow \pi_n(E) \to \pi_n(E, F) \longrightarrow \cdots$$

Also, we have $\pi_*: \pi_{n+1}(E, F) \to \pi_{n+1}(B, b_0)$. We claim that this map is an isomorphism. Once we know this, the exact sequence as claimed is derived immediately from the above exact sequence of the pair.

We define an inverse to $\pi_*: \pi_{n+1}(E, F) \to \pi_{n+1}(B, b_0)$.

Given $f: (I^n, \partial I^n) \to (B, b_0)$, there is a lifting

$$\begin{array}{ccc} & & E \\ & \overset{g}{\nearrow} & \downarrow \pi \\ I^n & \xrightarrow{f} & B \end{array}$$

The reason is that, since I^n is contractible, the map f is homotopic to the constant map $I^n \to b_0$. (The homotopy must in general deform ∂I^n off of b_0.) By the homotopy lifting property, any map, homotopic to a map which lifts, itself lifts. Clearly, $g: (I^n, \partial I^n) \to (E, F)$ determines an element in $\pi_n(E, F)$ which projects via π_* to f. This proves that π_* is onto.

To prove that it is one-to-one, we show that if $g_0, g_1: (I^n, \partial I^n) \to (E, F)$ are such that $\pi \circ g_0$ is homotopic to $\pi \circ g_1$ as maps $(I^n, \partial I^n) \to (B, b_0)$, then g_0 is homotopic to g_1. By homotopy lifting, we can assume that $\pi \circ g_0 = \pi \circ g_1$. The result now follows from the next lemma by letting $X = I^{n-1}$. ■

Lemma 4.12. *Given* $H: X \times I \to B$ *and two liftings* $\tilde{H}_1$ *and* $\tilde{H}_2: X \times I \to E$ *which agree in* $X \times \{0\}$*, there is a homotopy of lifting of H connecting* $\tilde{H}_1$ *and* $\tilde{H}_2$ *all of which agree on* $X \times \{0\}$.

Proof. Let $\tilde{h} = \tilde{H}|X \times \{0\}$. We have a commutative diagram

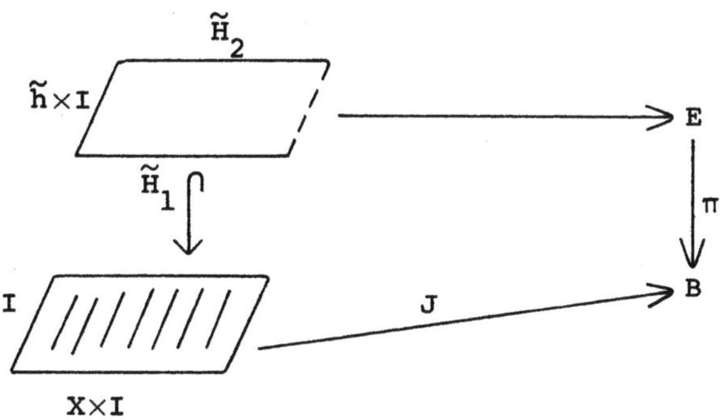

where the map $J: X \times I \times I \to B$ is projection onto $X \times I$ followed by H. Since $X \times I \times I$ deforms onto $X \times \{0\} \times I \cup X \times I \times \{0\} \cup X \times \{1\} \times I$, we can extend $\tilde{H}_1 \cup \tilde{h} \times I \cup \tilde{H}_2$ to a lifting of J. ■

4.7 Applications of the Exact Homotopy Sequence

1. $\pi_2(\mathbb{C}P^n) \cong \mathbb{Z}$ and $\pi_i(\mathbb{C}P^n) = \pi_i(S^{2n+1})$ for $i \neq 2$.
   This follows immediately from the homotopy long exact sequence of the Hopf fibration

$$S^1 \to S^{2n+1} \longrightarrow \mathbb{C}P^n$$

   and the fact that $\pi_i(S^1) = 0$ for $i > 1$. [To calculate the higher homotopy groups of S^1, recall that $\mathbb{R}^1 \xrightarrow{\exp} S^1$ is the universal cover. In general, the unique path lifting property of $\tilde{X} \to X$ implies that $\pi_i(\tilde{X}) \cong \pi_i(X)$ if $i > 1$ if $\tilde{X}$ is a covering space of X. Since $\mathbb{R}^1$ is contractible, $\pi_i(S^1) \cong \pi_i(\mathbb{R}^1) = 0$ for $i > 1$.]

2. $\pi_3(S^2) \cong \mathbb{Z}$.
   This is the special case of $n = 1$ in the above example.
3. Let ΩB be the loop space on B. It is the fiber of $\mathscr{P}B \to B$ where $\mathscr{P}B$ is the path space of B. Since $\mathscr{P}B$ is contractible, this gives $\pi_{i-1}(\Omega B) \cong \pi_i(B)$.

Chapter 5
Spectral Sequence of a Fibration

5.1 Introduction

We begin with a fibration

$$\begin{array}{ccc} F & \longrightarrow & E \\ & & \downarrow{\scriptstyle \pi} \\ & & B \end{array}$$

In Chap. 4, we saw that the homotopy groups of F,E,B are related by an exact sequence:

$$\ldots \longrightarrow \pi_n(F) \longrightarrow \pi_n(E) \longrightarrow \pi_n(B) \xrightarrow{\partial} \pi_{n-1}(F) \longrightarrow \ldots$$

Our goal now is to understand how the cohomologies of F, E, B are related. It is to be expected that the relationship will be somewhat complicated, because even in the case of a product $E = F \times B$, the Kunneth theorem giving $H^*(E)$ in terms of $H^*(F)$ and $H^*(B)$.

$$H^*(F \times B; \mathbb{Q}) = H^*(F; \mathbb{Q}) \otimes H^*(B; \mathbb{Q}).$$

is more complicated than the simple formula

$$\pi_n(F \times B) \cong \pi_n(F) \oplus \pi_n(B)$$

for the homotopy groups.

P. Griffiths and J. Morgan, *Rational Homotopy Theory and Differential Forms*,
Progress in Mathematics 16, DOI 10.1007/978-1-4614-8468-4_5,
© Springer Science+Business Media New York 2013

5.2 Fibrations over a Cell

The total space of a fibration over a CW complex is filtered by the increasing sequence of subspaces lying over the various skeleta of the base. To calculate the cohomology of a pair of spaces, one has a long exact sequence relating the cohomology of the subspace, the cohomology of the pair, and the cohomology of the total space. For a more general filtration, there is a much more complicated algebraic formalism—a spectral sequence—which relates the cohomology of the successive pairs in the filtration and the cohomology of the space.

Suppose that B is a connected CW complex with p-skeleton $B^{(p)}$, and let $E^{(p)} = \pi^{-1}(B^{(p)})$. Let $b_0 \in B$ be the base point and $F = \pi^{-1}(b_0)$.

Proposition 5.1. *If* B *is path connected and if* $\pi_1(B, b_0)$ *acts trivially on* $H^*(F)$, *then there are isomorphisms:*

$$\begin{aligned} H^n(E^{(p)}, E^{(p-1)}) &\cong \prod_{\{p-\text{cells in } B\}} H^n(\pi^{-1}(e^p), \pi^{-1}(\partial e^p)) \\ &\cong C^p(B; H^{n-p}(F)). \end{aligned}$$

Proof. Consider

$$\coprod_{\{p-\text{cells in } B\}} e^p \xrightarrow{\varphi} B$$

to be the map of all p-cells into B. Form $\varphi^*E \to \amalg e^p$ the induced fibration.Since $(\amalg e^p, \amalg \partial e^p) \to (B^{(p)}, B^{(p-1)})$ is a relative homeomorphism, so is $(\varphi^*E, \varphi^*E \amalg \partial e^p) \to (E^{(p)}, E^{(p-1)})$. Thus, by excision,

$$\begin{aligned} H^*(E^{(p)}, E^{(p-1)}) &\cong H^*(\varphi^*E, \varphi^*E|\partial e^p) \\ &= \prod_{p-\text{cells}} H^*(\varphi^*E|e^p, \varphi^*E|\partial e^p). \end{aligned}$$

■

Consider a fibration $E^1 \xrightarrow{\pi} e^p$. Let $F_0 = \pi^{-1}(0)$. We have a diagram

$$\begin{array}{ccc} F_0 \times \{0\} & \hookrightarrow & E^1 \\ \downarrow & & \downarrow \pi \\ F_0 \times e^p & \xrightarrow{p_2} & e^p \end{array}$$

Since e^p is contractible, the map p_2 lifts to a map $\tilde{p}\colon F_0 \times e^p \to E^1$, extending the inclusion of F_0 into E^1. This gives a fiberwise map $\tilde{p}\colon (F_0 \times e^p, F_0 \times \partial e^p) \to (E^1, E^1|\partial e^p)$. Comparing the homotopy long exact sequences, we see that $\tilde{p}$ induces an isomorphism on the relative homotopy groups. Thus, it is a homotopy equivalence of pairs.

Thus, in our case, we have

$$H^*(\varphi^*E|e^p, \varphi^*E|\partial\, e^p) \;\cong H^*(F_0) \otimes H^*(e^p, \partial e^p).$$

If $\pi_1(B, b_0)$ acts trivially on $H^*(F)$, then choosing a path from b_0 to $0 \in e^p \subset B^{(p)}$ gives an identification of $H^*(F)$ with $H^*(F_0)$ which is independent of the path. Thus, we can identify $H^*(F_0) \otimes H^*(e^p, \partial e^p)$ with $H^*(F) \otimes H^*(e^p, \partial e^p)$. This gives

$$\begin{aligned} H^*(E^{(p)}, E^{(p-1)}) \;&\cong \prod_{p-\text{cells}} H^*(e^p, \partial e^p) \otimes H^*(F) \\ &\cong C^p(B; H^*(F)). \end{aligned}$$

Returning now to our proposition, we may think of $H^*(E^{(p)}, E^{(p-1)})$ as a first approximation to $H^*(E^{(p)})$, of $H^*(E^{(p)}; E^{(p-2)})$ as a second and somewhat better approximation, of $H^*(E^{(p)}; E^{(p-3)})$ as a third approximation, etc. It is pretty clear that the successive approximations are related by exact sequences in which the missing terms are given by the previous proposition. Thus, we can expect to get a whole chain of exact sequences, each related to the previous one and each one giving us better and better approximation to $H^*(E)$. The algebra involved is formalized in the following.

5.3 Generalities on Spectral Sequences

Definition. A (first quadrant) *spectral sequence* consists of abelian groups $E_r = \{E_r^{p,q}\}_{p,q\geq 0} (r \geq 0)$ and maps $d_r\colon E_r^{p,q} \to E_r^{p+r,q-r+1}$ with $d_r^2 = 0$, such that the homology $H(E_r, d_r) = E_{r+1}$. In detail,

$$E_{r+1}^{p,q} = \frac{\ker d_r\colon E_r^{p,q} \to E_r^{p+r,q-r+1}}{\operatorname{im} d_r\colon E_r^{p-r,q+r-1} \to E_r^{p,q}}.$$

Remark. The most important thing about spectral sequences is to have in mind the picture: For each term E_r, we plot the first quadrant in the (p, q) plane and put in the group $E_r^{p,q}$ at the lattice point (p, q). Then the differentials map according to the picture.

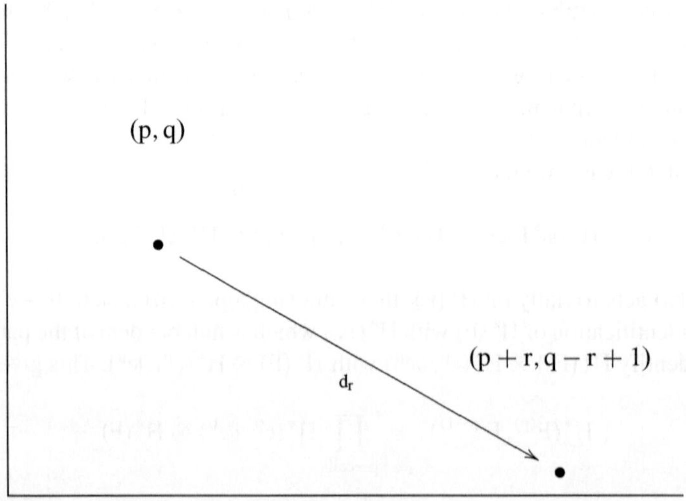

An element $\alpha \in E_r^{p,q}$ is said to *live to infinity* if $d_r\alpha = 0$ (and thus α defines an element in $E_{r+1}^{p,q}$), $d_{r+1}\alpha = 0$, …, etc. An element $\beta \in E_r^{p,q}$ is said to be *killed* if $d_r\beta = \ldots = d_{s-1}\beta = 0$, but $\beta = d_s\gamma$ for some $\gamma \in E^{p-s,q+s-1}$. The spectral sequence is said to degenerate at E_r if $E_r \cong E_{r+1} \cong \ldots$ etc.; it is said to be *degenerate* if $E_2 \cong E_3 \cong E_4 \cong \cdots$ Spectral sequences are a necessary and useful tool and should not be either over- or underestimated. As with evaluation of calculus integrals, they are best learned by practice.

As an exercise in understanding pictures, we prove the following lemma:

Lemma 5.2. *In a first quadrant spectral sequence if* $N > p, q+1$, *then* $E_N^{p,q} \cong E_{N+1}^{p,q}$.

Proof. In the picture, we find

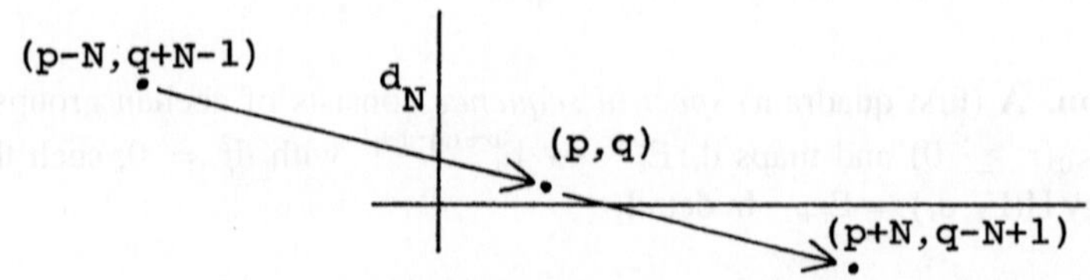

so that $d_N \equiv 0$ on $E_N^{p,q}$ and $E_N^{p-N,q+N-1} = 0$ so that the image of d_N in $E_N^{p,q} = 0$. ■

Definition. We set $E_\infty^{p,q} = E_N^{p,q}$ for $N > p,\ q+1$ and $E_\infty^n = \oplus_{p+q=n} E_\infty^{p,q}$.

As another exercise in the definitions, we have the following:

Lemma 5.3. *Given a first quadrant spectral sequence* $\{E_r^{p,q}\}$, *we have*

$$E_2^{1,0} = E_\infty^{1,0}.$$

We also have an exact sequence

$$\{0\} \longrightarrow E_2^{0,1} \xrightarrow{d_2} E_2^{2,0} \longrightarrow E_\infty^{2,0} \longrightarrow \{0\}.$$

Proof. The picture is

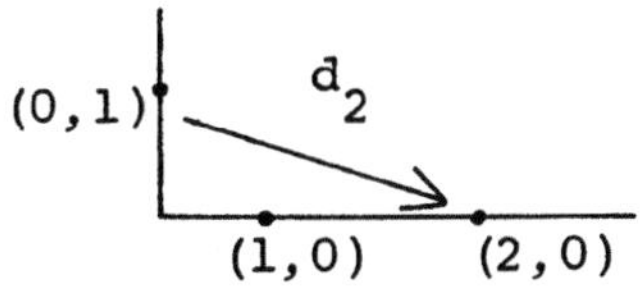

■

5.4 The Leray–Serre Spectral Sequence of a Fibration

The main result concerning spectral sequences which interests us is the following:

Theorem 5.4 (Leray–Serre). *Let* $F \to E \to B$ *be a fibration in which* B *is a connected CW complex with* $\pi_1(B)$*acting trivially on the cohomology of the fiber. Then there is a spectral sequence* $\{E_r^{p,q}\}$ *in which*

$$E_1^{p,q} = C^p(B, H^q(F)) \text{ and } d_1 = \text{coboundary map of } C^*(B, H^q(F));$$

$$E_2^{p,q} = H^p(B, H^q(F)); \text{ and}$$

$$E_N^{p,q} = \frac{\ker\{H^{p+q}(E) \to H^{p+q}(E^{(p-1)})\}}{\ker\{H^{p+q}(E) \to H^{p+q}(E^{(p)})\}}, \; N > p,\; q+1.$$

Remarks. If we define

$$\mathscr{F}^p H^{p+q}(E) = \ker\{H^{p+q}(E) \to H^{p+q}(E^{(p-1)})\}$$

then clearly

$$H^n(E) = \mathscr{F}^0 H^n(E) \supset \mathscr{F}^1 H^n(E) \supset \ldots \supset \mathscr{F}^n H^n(E) \supset \mathscr{F}^{n+1} H^n(E) = 0,$$

so that the groups $\mathscr{F}^p H^n(E)$ give a decreasing filtration on $H^n(E)$ whose associated graded abelian group is $\oplus_{p+q=n} E_\infty^{p,q}$. This is usually written as

$$E_r^{p,q} \Longrightarrow H^{p+q}(E)$$

and one says that *the spectral sequence converges (or abuts)* to $H^*(E)$.

Proof. Let $\mathrm{Sing}^*(E)$ be the singular cochains of E. We define a filtration

$$F^p(\mathrm{Sing}^*(E)) = \ker\{\mathrm{Sing}^*(E) \longrightarrow \mathrm{Sing}^*(E^{(p-1)})\}.$$

It is clear that $\delta F^p \subset F^p$. Thus, $Sing^*(E)$ is a cochain complex with a decreasing filtration preserved by δ. Under this circumstance, it is a purely algebraic result that there is a spectral sequence which converges to the cohomology on $Sing^*(E)$. The terms are

$$\begin{aligned} E_r^{p,q} &= Z_r^{p,q}/B_r^{p,q} \\ Z_r^{p,q} &= \{\text{cochains } \mu \text{ in } F^p(Sing^{p+q}(E)) \text{ such that} \\ &\qquad \delta\mu \in F^{p+r}(Sing^{p+q+1}(E))\} \\ B_r^{p,q} &= \{Z_r^{p,q} \cap F^{p+1} + \delta F^{p-r+1}(Sing^{p+q-1}(E)) \cap F^p\}. \end{aligned}$$

It is a computational result that δ induces a differential, denoted d_r on E_r, and that $H^*(E_r)$ (computed using the differential d_r) is E_{r+1}. By definition $E_0^{p,q}$ is $F^p(Sing^{p+q}(E))/F^{p+1}(Sing^{p+q}(E)) \cong Sing^{p+q}(E^{(p)}, E^{(p-1)})$. The differential d_0 is the usual singular coboundary map on the relative cochains. Thus, $E_1^{p,q} = H^{p+q}(E^{(p)}, E^{(p-1)})$. Proposition 5.1 gives $E_1^{p,q} \cong C^p(B; H^q(F))$. It is easy to see that under this identification, d_1 becomes $\delta_B \otimes 1$. Thus, $E_2^{p,q} = H^p(B; H^q(F))$. This completes the proof of the theorem. ∎

There is another version of the Leray–Serre spectral sequence for C^∞-manifolds. Let $\pi: E \to B$ be a C^∞ map between C^∞ manifolds which is locally (in B) differentiably equivalent to a projection $F \times U \to U$. (We call such a smoothly locally trivial fibration.) We define a filtration on the C^∞-differential forms of E, denoted $\Omega^*(E)$. The condition that ω be in $F^i(\Omega^*(E))$ is a pointwise condition which is required to hold for each point. A form ω^ℓ is contained in F^i if and only if for every $p \in E$ and every set of tangent vectors $\tau_1(p, \ldots, \tau_\ell(p$ we have

$$< \omega^\ell(p), \tau_1(p) \wedge \ldots \wedge \tau_\ell(p) >= 0$$

whenever $\ell - i + 1$ of the tangent vectors $\tau_j(p)$ are vertical, i.e., are in the kernel of $D\pi_p$.

Explanation: ω^ℓ is an ℓ-form, and hence, $\omega^\ell(p) \in \wedge^\ell TE_p^*$. Thus, if $\tau_1(p), \ldots, \tau_\ell(p)$ are tangent vectors to E at p, then $< \omega^\ell(p), (\tau_1(p) \wedge \ldots \wedge \tau_\ell(p)) >$ is a real number.

Claim.

$$E_0^{p,q} = \frac{F^p(\Omega^{p+q}(E))}{F^{p+1}(\Omega^{p+q}(E))}$$

is identified with C^∞ p-forms on B with values in C^∞ q-forms on the fibers.

Proof of Claim. Let $\omega^{p+q} \in F^p(\Omega^{p+q}(E))$. Let $\tau_1(b), \ldots, \tau_p(b)$ be tangent vectors to B at $b = \pi(p)$. Let $\tilde{\tau}_i$ be a tangent vector to E at p which projects onto $\tau_i(b)$. Let $\tilde{\tau}_{p+1}, \ldots, \tilde{\tau}_{p+q}$ be tangent vectors to $\pi^{-1}(p) = F_p \subset E$. Then $< \omega^{p+q}(p), \tilde{\tau}_1 \wedge$

$\ldots \wedge \tilde{\tau}_{p+q} >$ is independent of the liftings $\tilde{\tau}_1, \ldots \tilde{\tau}_p$ of the tangent vectors in B. This defines $F^p(\Omega^{p+q})$ to the space of p-forms on B with values in q-forms on the fibers. Clearly, this map factors through $F^p(\Omega^{p+q})/F^{p+1}(\Omega^{p+q})$. One checks easily that it induces an isomorphism

$$F^p(\Omega^{p+q})/F^{p+1}(\Omega^{p+q}) \cong \text{(p-forms on B with values in q-forms on the fibers.}$$

The differential d_0 becomes differentiation in the fiber direction under this identification. Hence,

$$E_1^{p,q} = \Omega^p(B; H^q(F));$$

i.e., $E_1^{p,q}$ is p-forms on B with coefficients in the vector bundle of qth deRham cohomology along the fibers. The vector bundle of deRham cohomology along the fiber has a natural product structured locally (since deRham cohomology is identified with singular cohomology with real coefficients which is locally constant). That is to say the bundle of deRham cohomology along the fibers has a natural flat connection (called the Gauss–Manin connection). The condition that the fundamental group of the base acts trivially on the cohomology of the fiber is the condition that this connection is trivial, so that this bundle of deRham cohomology is identified with the constant coefficients of deRham cohomology of the fiber over the base point. One then shows that d_1 becomes $d_B \otimes 1$, i.e., differentiation in the base direction. Consequently,

$$E_2^{p,q} = H^p(B; H^q(F)) \quad \text{(deRham cohomology).}$$

This is the C^∞-version, or deRham version, of the Leray–Serre spectral sequence.

The last description we give of this spectral sequence is the one originally given by Serre in [21]. One considers the singular cubical cochains $Q^*(E)$. The nth chain group is defined by taking the free abelian group generated by maps of the n-cube, I^n into E and setting equal to 0 the degenerate cubes, i.e., those that factor through $I^n \to I^{n-1}$ defined by

$$(t_1, \ldots, t_n) \mapsto (t_1, \ldots, t_{n-1}).$$

The cochains are dual to these chain groups. The boundary map is defined by restricting to the various codimension-1 faces with signs. These again calculate the usual singular cohomology. We say that $\alpha \in Q^n(E)$ is in filtration level i, $\alpha \in F^i$, if α vanishes when evaluated on any cubical chain $\varphi: I^n \to E$ which has the property that $\pi \circ \varphi: I^n \to B$ factors through the projection $I^n \to I^{i-1}$, $(t_1, \ldots, t_n) \to (t_1, \ldots, t_{i-1})$. Assuming that the fundamental group of B acts trivially on the cohomology of the fiber $E_1^{p,q} = Q^p(B) \otimes H^q(F)$, d_1 is identified with the usual coboundary map on Q^*. Since the non-degenerate cubical cochain complex calculates singular cohomology, one finds that $E_2^{p,q} = H^p(B; H^q(F))$ (with $\mathbb{Z}$-coefficients). ■

Remarks. (i) There is also a homology spectral sequence for a fibration, $\{E^r_{p,q}\}$, with $E^2_{p,q} = H_p(B; H_q(F))$. Over $\mathbb{Q}$ the two spectral sequences are dual in the obvious sense.

(ii) If $\pi_1(B, b_0)$ acts nontrivially on $H^*(F)$, then there is still a spectral sequence. In this case $E_2^{p,q}$ is $H^p(B;\ H^q(F))$ where the coefficients are the locally trivial, but globally nontrivial coefficient system on the base which is given by the flat connection determined by the action of the fundamental group of the base on the cohomology of the fiber.

(iii) There is a ring structure on the terms in the cohomology spectral sequence such that the differentials d_i are derivations. In particular, if we use, say $\mathbb{Q}$-coefficients, and assume the action of the fundamental group of the base on the cohomology of the fiber is trivial, then

$$E_2^{p,q} \cong H^p(B) \otimes H^q(F)$$
$$\cong E_2^{p,0} \otimes E_2^{0,q} \text{and}$$
$$d_2^{(p,q)} = d_2^{(p,0)} \otimes \mathrm{Id} + (-1)^p \mathrm{Id} \otimes d_2^{(0,q)} = (-1)^p \mathrm{Id} \otimes d_2^{(0,q)}$$
$$\text{since } d_2^{(p,0)} = 0.$$

This is proved by using the usual cup product on cochains and being careful about the formulae; for more details, cf. [23] pp. 490–498. If one is willing to work over $\mathbb{R}$, then this multiplicative property is easily seen using the differential form version of the spectral sequence discussed above. It should be remarked that the most sophisticated computations involving spectral sequences seem to use the ring structure and derivation property.

5.5 Examples

1. **Complex Projective Space.** We shall compute the cohomology ring of $\mathbb{C}P^n = \mathbb{P}^n$ using the *Hopf fibration* $S^1 \to S^{2n+1} \to \mathbb{P}^n$.

Using $\mathbb{Z}$-coefficients, the E_2 term is $H^*(S^1) \otimes H^*(\mathbb{C}P^n)$ and $E_\infty \Rightarrow H^*(S^{2n+1})$. Since $H^q(S^1) = 0$ for $q > 1$, $E_2^{p,q} = 0$ for $q > 1$ and so $d_3 = d_4 = \ldots = 0$ and $E_3 \cong E_4 \cong \ldots \cong E_\infty$. Thus, all the "action" in the spectral sequence occurs at the E_2 level, and in particular $E_\infty^{p,q} = E_3^{p,q} = 0$ unless $p + q = 0,\ 2n + 1$. The picture of E_2 is

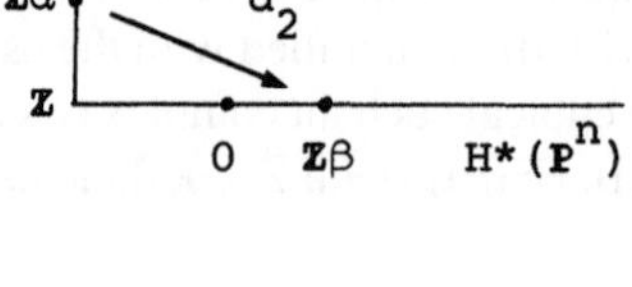

where $\alpha \in E_2^{0,1} \cong H^1(S^1)$ is a generator.

To begin with, $E_2^{1,0} = 0$ since $E_\infty^{1,0} \cong E_2^{1,0}$. Next the map

$$E_2^{0,1} \xrightarrow{d_2} E_2^{2,0}$$

is an isomorphism since $E_3^{p,q} = 0$ for $1 \leq p + q \leq 2$. Set $\beta = d_2\alpha$ so that $E_2^{2,1} \cong \mathbb{Z}(\alpha \otimes \beta)$. Then $d_2(\alpha \otimes \beta) = -\beta^2 \in E_2^{4,0} \cong H^4(\mathbb{P}^n)$ since d_2 is a derivation. Continuing in this way, we see that

$$\begin{aligned} H^{2q}(\mathbb{P}^n) &= \mathbb{Z}[\beta^q] \quad (q \leq n) \\ H^{2q+1}(\mathbb{P}^n) &= 0 \end{aligned}$$

So that $H^*(\mathbb{P}^n) \cong \mathbb{Z}[\beta]/(\beta^{n+1})$; i.e., $H^*(\mathbb{P}^n)$ it is a truncated polynomial algebra.

2. **Rational Cohomology of** $K(\mathbb{Z}, n)$

We shall show in Chap. 7 that there is a space, unique up to homotopy equivalence, which has only one nonzero homotopy group, that being π_n, and that group is isomorphic to $\mathbb{Z}$. Such a space is denoted $K(\mathbb{Z}, n)$. We shall prove the fundamental result here that

$$H^*(K(\mathbb{Z}, 2k); \mathbb{Q}) \cong \mathbb{Q}[\alpha],$$

and

$$H^*(K(\mathbb{Z}, 2k+1); \mathbb{Q}) \cong \mathbb{Q}(\beta).$$

(The first is a polynomial algebra and the second is an exterior algebra.) The proof of these results is by induction on n. For $n = 1$, $K(\mathbb{Z}, 1) \cong S^1$ and the result is immediate. Consider the inductive step from $(2k-1)$ to $2k$. We study the path fibration over $K(\mathbb{Z}, 2k)$:

$$\begin{array}{ccc} K(\mathbb{Z}, 2k-1) & \longrightarrow & \mathcal{P}(K(\mathbb{Z}, 2k)) \\ & & \downarrow \\ & & K(\mathbb{Z}, 2k) \end{array}$$

Since $\mathscr{P}$ is contractible, the fiber is $K(\mathbb{Z}, 2k-1)$ and $E_\infty^{p,q} = 0$ if $(p, q) \neq (0, 0)$. Invoking the inductive hypothesis, we see that the E_2-term looks like

$H^*(K(\mathbb{Z}, 2k); \mathbb{Q})$	$(*, 2k-1)$
.	
.	
.	
.	
$H^*(K(\mathbb{Z}, 2k); \mathbb{Q})$	$(*, 0)$

Thus, $E_2 = E_3 = \ldots = E_{2k-1}$ and $E_{2k} = E_\infty$. First note that

$$H^i(K(\mathbb{Z}, 2k); \mathbb{Q}) = \begin{cases} 0, & i < 2k \\ \mathbb{Q}, & i = 2k \end{cases}$$

(This is a consequence also of the Hurewicz theorem and the universal coefficient theorem). Let β be a generator of $H^{2k-1}(K(\mathbb{Z}, 2k-1); \mathbb{Q})$, and let $\alpha = d_{2k}(\beta)$. Then α is a generator of $H^{2k}(K(\mathbb{Z}, 2k); \mathbb{Q})$. Suppose inductively that we have shown that $\mathbb{Q}[\alpha] \to H^*(K(\mathbb{Z}, 2k); \mathbb{Q})$ is an isomorphism for all degrees $\leq t(2k)$. Then we have

$E_{2k} =$

0	0	0	0	0	0...	0
β	0	$\alpha \otimes \beta$	0	$\alpha^2 \otimes \beta$	0...	$\alpha^t \otimes \beta$
0	0	0	0	0	0...	0
0	0	0	0	0	0...	0
0	0	0	0	0	0...	0
1	0	α	0	α^2	0...	α^t

It results from the fact that $E_{2k+1}^{p,q} = 0$ for all $(p, q) \neq (0, 0)$, that $H^i(K(\mathbb{Z}, 2k); \mathbb{Q}) = 0$ for $2kt < i < 2k(t+1)$, and that $H^{2k(t+1)}(K(\mathbb{Z}, 2k); \mathbb{Q})$ is isomorphic to $\mathbb{Q}$ and generated by $d_{2k}(\beta \otimes \alpha^t)$. By the derivation property, $d_{2k}(\beta \otimes \alpha^t) = \alpha \otimes \alpha^t = \alpha^{t+1}$. This completes the induction step $(2k-1) \to 2k$.

For the inductive step $2k \to 2k+1$, we use the following path fibration:

$$\begin{array}{ccc} K(\mathbb{Z}, 2k) & \longrightarrow & \mathcal{P}(K(\mathbb{Z}, 2k+1)) \\ & & \downarrow \\ & & K(\mathbb{Z}, 2k+1). \end{array}$$

Here E_2 is

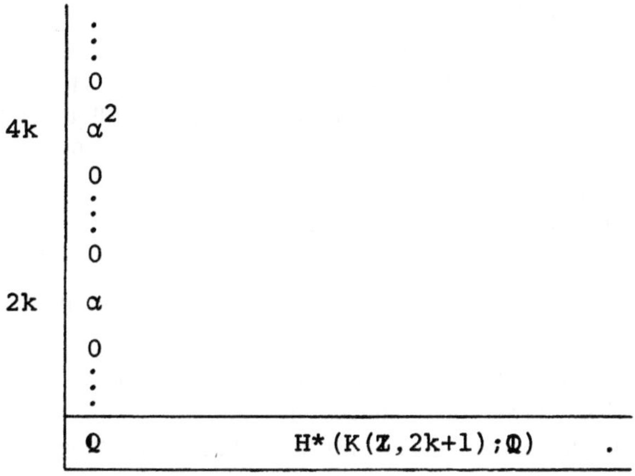

An argument similar to the one in the other case shows that

$$\begin{cases} E_2 &= \ldots = E_{2k+1} \\ E_{2k+2} &= \ldots = E_\infty \\ E_\infty^{p,q} &= 0 \text{ for } p+q > 0. \end{cases}$$

It follows that $E_{2k+1}^{0,2k} \xrightarrow{d_{2k+1}} E_{2k+1}^{2k+1,0}$ is an isomorphism, and so we set $\beta = d_{2k+1}\alpha$. By the derivation property, $d_{2k+1}(\alpha^2) = 2\,\alpha \otimes d_{2k+1}\alpha = 2\,\alpha \otimes \beta$, and so

$$E_{2k+1}^{0,4k} \xrightarrow{d_{2k+1}} E_{2k+1}^{2k+1,2k}$$

is an isomorphism (this is where we need $\mathbb{Q}$-coefficients). This gives that $H^{4k+2}(K(\mathbb{Z}, 2k+1), \mathbb{Q}) = 0$, and continuing on we obtain the desired result.

3. **Grassmannians.** Let $G(k, N)$ be the Grassmann manifold of complex k-planes in $\mathbb{C}^N$ and $G(k) = \lim_{N\to\infty} G(k, N)$ the *infinite Grassmanian*. Using $\mathbb{Z}$ coefficients, we want to prove that

$$H^p(G(k)) \cong \mathbb{Z}\,[c_1, c_2, \ldots, c_k],\ \ c_j \in H^{2j}(G(k)).$$

When $k = 1$, we have $G(1) = \mathbb{C}P^\infty = K(\mathbb{Z}, 2)$ and we have proved the result in this case. Assuming the result for $k-1$, we consider the two fibrations

$$\begin{array}{ccccc} & & S^{2k-1} & & \\ & & \downarrow & & \\ G(k-1) & \longleftarrow & F & \longleftarrow & S^\infty \\ & & \downarrow & & \\ & & G(k) & & \end{array}$$

where $F = \{(v, \pi) \mid \pi$ is a k-plane in $\mathbb{C}^\infty$ and $v \in \pi$ is a unit vector$\}$. The map $F \to G(k-1)$ is given by sending (v, π) to the $(k-1)$ plane in π perpendicular to v. Clearly, $F \to G(k-1)$ is a fibration with fiber S^∞ which is contractible. Hence, $F \cong G(k-1)$. The fibration $F \to G(k)$ is given by sending (v, π) to π. The fiber here is S^{2k-1}. Thus, we have a spectral sequence such that $E_2^{p,q} = H^p(G(k); H^q(S^{2k-1}))$ and E_∞ converges to $H^*(G(k-1))$. It must be the case that $E_2 = E_3 = \ldots = E_{2k-1}$ and that $E_{2k} = E_\infty$. Let c_k be $d_{2k}(\alpha)$ where α is the generator of $H^{2k-1}(S^{2k-1})$. From the spectral sequence, we have exact sequences

$$H^{i+2k-1}(G(k-1)) \longrightarrow H^i(G(k)) \xrightarrow{\cup c_k} H^{i+2k}(G(k)) \longrightarrow H^{i+2k}(G(k-1)).$$

By induction $H^*(G(k-1))$ is a polynomial algebra generated by $c_1, \ldots, c_{k-1}$. Also the inclusion $G(k-1) \hookrightarrow G(k)$ induces an isomorphism on H^i for $i < 2k-1$. This implies that the classes $c_1, \ldots, c_{k-1}$ lift to $H^*(G(k))$. It follows that $H^*(G(k)) \to H^*(G(k-1))$ is onto. Consequently, the above sequence becomes

$$0 \longrightarrow H^i(G(K)) \xrightarrow{\cup c_k} H^{i+2k}(G(k)) \longrightarrow H^{i+2k}(G(k-1)) \longrightarrow 0.$$

Using this one proves easily that $H^*(G(k))$ is a polynomial algebra on $c_1, \ldots, c_k$.

Chapter 6
Obstruction Theory

6.1 Introduction

Recall, in the proof of the Whitehead theorem, we showed that if X is a CW complex and Y is a space with $\pi_i(Y) = 0$ for all $i \geq 0$, then any map $f: X \to Y$ is homotopic to the base point map $X \to y_0 \in Y$. The proof was by induction over the skeleta of X. Obstruction theory is a generalization of this technique to the case when the homotopy groups of Y are not necessarily zero. It does not give a complete understanding of when a map $f: X \to Y$ is homotopic to a constant, or more generally when two maps $f_1, f_2: X \to Y$ are homotopic, but it gives some insight.

Obstruction theory not only deals with the uniqueness question "when are two maps f_1, $f_2: X \to Y$ homotopic?"; it also deals with the existence question of constructing maps from X to Y. In both cases, we work inductively over the skeleta. In the first, we suppose that we have a homotopy h from $f_1|X^{(n)}$ to $f_2|X^{(n)}$

$$h: X^{(n)} \times I \to Y$$

and we ask if it extends to $X^{(n+1)} \times I \to Y$. In the second, we suppose that we have a map $f: X^{(n)} \to Y$, and we ask if it extends to a map $\hat{f}: X^{(n+1)} \to Y$.

We will deal with the second case first.

Lemma 6.1. *Suppose that* Y *is simply connected.*

1. *There is a natural bijection*

$$\pi_n(Y) \longleftrightarrow [S^n, Y],$$

where $[S^n, Y]$ *means the free homotopy classes of maps of* S^n *into* Y *(with no restrictions on the image of base point).*

P. Griffiths and J. Morgan, *Rational Homotopy Theory and Differential Forms*, Progress in Mathematics 16, DOI 10.1007/978-1-4614-8468-4_6,
© Springer Science+Business Media New York 2013

2. *Let* U *be the complement in* S^{n+1} *of the interiors of a finite disjoint collection of disks* $\cup_i D_i^{n+1}$, *and denote the boundary of* D_i^{n+1} *by* S_i^n. *If a finite collection of maps* $f_i: S_i^n \to Y$ *extends over* U, *then* $\sum_i [f_i] = 0$ *in* $\pi_n(Y)$.

Proof. We consider the first statement. There is always a function

$$\pi_n(Y) \longrightarrow [S^n, Y]$$

obtained by ignoring the base points. If $\pi_0(Y) = 0$, then it is onto. This follows by applying the homotopy extension property (h. e. p.) to $S^n \times \{0\} \cup * \times I \subset S^n \times I$ (* is the base point of S^n). If in addition $\pi_1(Y) = 0$, then the map is 1–1. This follows by applying h. e. p. to $(S^n \times \{0 \cup 1\} \cup * \times I) \times I$ in $(S^n \times I) \times I$.

For the second statement, let $F: U \to Y$ be an extension of the f_i. There is a finite disjoint union of arcs α_j in U, with end points in the boundary of U such that the complement in U of an open regular neighborhood of the union of these arcs is an $(n + 1)$-disk D'. We deform F so that the closure of the neighborhood of these arcs maps to the base point of Y. Then we see that $\sum_i [f_i] = [F|_{\partial D'}]$ in $\pi_n(Y)$. The latter is obviously the trivial element. ■

6.2 Definition and Properties of the Obstruction Cocycle

We assume for the rest of this chapter that Y is simply connected.

Let (X, A) be a CW-pair. Suppose given $f_n: X^{(n)} \cup A \to Y$. We define the obstruction cochain $\tilde{\mathcal{O}}(f_n) \in C^{n+1}(X, A; \pi_n(Y))$ as follows: If e_α^{n+1} is an oriented (n+1)-cell of (X, A), then its attaching map $c_\alpha: S^n \to X^{(n)} \cup A$ composed with f gives $f \circ c_\alpha: S^n \to Y$, which determines an element of $\pi_n(Y)$. If we reverse the orientation on e_α^{n+1} (and hence on ∂e_α^{n+1}), then the resulting element in $\pi_n(Y)$ changes sign. Thus, there is a well-defined homomorphism $C_{n+1}(X, A) \to \pi_n(Y)$. We denote it by $\tilde{\mathcal{O}}(f_n)$ and call it the *obstruction cochain.*

Since Y is simply connected, the obstruction cochain groups are trivial for $n = 0, 1$. From now on we fix orientations for each cell of (X, A) and use the symbol for the cell to represent this oriented cell.

Lemma 6.2. *The obstruction cocycle* $\tilde{\mathcal{O}}(f_n)$ *satisfies the following properties:*

(1) It is an invariant of the homotopy class of f_n.
(2) It is 0 if, and only if, f_n *extends to a map* $f_{n+1}: X^{(n+1)} \cup A \to Y$.
(3) It is a cocycle; i.e., $\delta\tilde{\mathcal{O}}(f_n): C_{n+2}(X, A) \to \pi_n(Y)$ *is* 0.
(4) If $g_n: X^{(n)} \cup A \to Y$ *agrees with* f_n *on* $X^{(n-1)} \cup A$, *then* $\tilde{\mathcal{O}}(g_n) - \tilde{\mathcal{O}}(f_n)$ *is a coboundary.*
(5) By varying the homotopy class of f_n *relative to* $X^{(n-1)} \cup A$, *we can change* $\tilde{\mathcal{O}}(f_n)$ *by an arbitrary coboundary.*

Proof. (1) and (2) are immediate from the definitions. Before proving (3), (4), and (5), we review a little of the terminology associated with cohomology and then prove a necessary lemma. ■

Let $\{C_*, \partial\}$ be a chain complex. Its homology is defined by

$$H_n(C) = \frac{\ker \partial_n}{\operatorname{Im} \partial_{n+1}} = \frac{\text{cycles}}{\text{boundaries}}.$$

If G is an arbitrary abelian group, then we define a cochain complex with coefficients in G by

$$\ldots \longleftarrow \operatorname{Hom}(C_n, G) \overset{\partial * = \delta}{\longleftarrow} \operatorname{Hom}(C_{n-1}, G) \longleftarrow \ldots.$$

Clearly, $\delta \circ \delta = 0$. We define

$$H^n(C; G) = \frac{\ker \delta_n}{\operatorname{Im} \delta_{n-1}} = \frac{\text{cocycles}}{\text{coboundaries}}.$$

From this description, it is easy to see that if σ_n is an n-cochain and τ_{n+1} is an $(n+1)-$ chain, then $< \delta\sigma_n, \tau_{n+1} > = < \sigma_n, \partial\tau_{n+1} >$. Thus, σ_n is a cocycle if, and only if, it annihilates all boundaries.

Proof of (3). $\delta(\tilde{\mathcal{O}}(f_n)) = 0$.

To prove this, we need to show that $< \tilde{\mathcal{O}}(f_n), \partial\gamma_{n+2} > = 0$ for all $(n+2)$-cells γ_{n+2}. Let $\sum_\alpha a_\alpha [e_\alpha]$ be the boundary of γ_{n+2} in the CW chain complex. Then the attaching map for γ, $\partial\gamma_{n+2}: S^{n+1} \to X^{(n+1)} \cup A$ is homotopic to a map $S^{n+1} \to X^{(n+1)} \cup A$ with the property that the preimage of the interior of each e_α is a disjoint union of disks mapping homeomorphically onto the interior of the cell and the sum of the orientation signs of these maps is a_α. It follows from Lemma 6.1 that $\sum_\alpha a_\alpha [\partial e_\alpha] = 0$ in $\pi_{n+1}(Y)$. Thus,

$$\begin{aligned}
< \tilde{\mathcal{O}}(f_n), \partial\gamma_{n+2} > &= < \tilde{\mathcal{O}}(f_n), \Sigma_\alpha a_\alpha [e_\alpha, \partial e_\alpha] > \\
&= \Sigma_\alpha a_\alpha < \tilde{\mathcal{O}}(f_n), [e_\alpha, \partial e_\alpha] > \\
&= \Sigma_\alpha a_\alpha [f_n \circ \partial e_\alpha] \\
&= (f_n)_* \Sigma_\alpha a_\alpha [\partial e_\alpha] \\
&= (f_n)_*(0) = 0.
\end{aligned}$$

■

Proof of (4). For each n-cell e^n_α in $X \setminus A$ we have $f_n|\partial e^n_\alpha = g_n|\partial e^n_\alpha$. Thus, we can form the difference element "$f_n - g_n$": $S^n \to Y$.

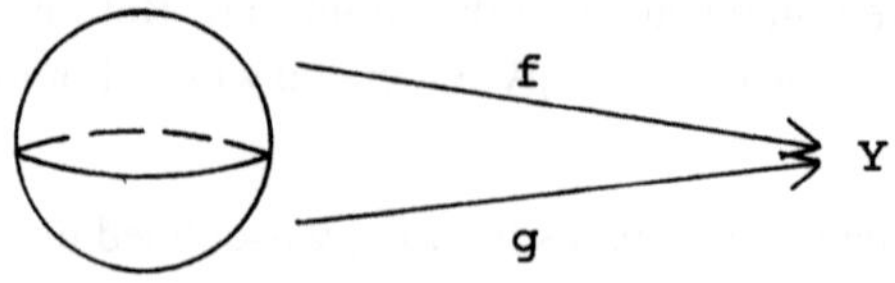

This defines a cochain "$f_n - g_n$": $C_n(X, A) \to \pi_n(Y)$. ■

claim. $\delta(\text{“}f_n - g_n\text{”}) = \tilde{\mathcal{O}}(f_n) - \tilde{\mathcal{O}}(g_n)$.

Proof. Let e_α^{n+1} be an (n+l)-cell of (X, A) with $\partial e_\alpha^{n+1} = \Sigma a_{\alpha\beta} e_\beta^n$ in $C_n(X, A)$. As before we can deform $\partial e_\alpha^{n+1}: S^n \to X^{(n)} \cup A$ by a homotopy to a map b_α with the property that the preimage of each n-cell, e_β under b_α is a finite number of n-disks each mapping homeomorphically to e_β and counted with signs coming from the orientations the number is $a_{\alpha\beta}$. The composites $f \circ b_\beta$ and $g \circ b_\beta$ agree on the complement of these disks. The "difference" element on each disk above e_β is $\pm$"$f_n - g_n$"(e_β), where, as before the sign is determined by the orientations. It is now an easy extension of the argument in Lemma 6.1 to show that

$$[f \circ \partial e_\alpha] - [g \circ \partial e_\alpha] = \sum_\beta a_{\alpha\beta} \text{“}f_n - g_n\text{”}(e_\beta).$$

Thus, we have

$$\begin{aligned} < \tilde{\mathcal{O}}(f_n), \partial e_\alpha^{n+1} > - < \tilde{\mathcal{O}}(g_n), \partial e_\alpha^{n+1} > &= f_n \circ \partial e_\alpha^{n+1} - g_n \circ \partial e_\alpha^{n+1} \\ &= \Sigma a_{\alpha\beta} < \text{“}f_n - g_n\text{”}, e_\beta^n > \\ &= < (\text{“}f_n - g_n\text{”}), \partial e_\alpha^{n+1} > \\ &= < \delta(\text{“}f_n - g_n\text{”}), e_\alpha^{n+1} > . \end{aligned}$$

■

Proof of (5). Let e_0^n be an n-cell of(X, A) and let g be an element of $\pi_n(Y)$. Define a cochain $\mu \in C^n(X, A; \pi_n(Y)$ by $< \mu, e_0^n >= g$ and $< \mu, e^n >= 0$ for all other n-cells e^n. We shall show how to change f_n to f'_n so that $\tilde{\mathcal{O}}(f_n) - \tilde{\mathcal{O}}(f'_n) = \delta(\mu)$. Since these cochains generate the entire cochain group, this will suffice to establish 5. Choose a small ball $B^n \subset \text{int}\, e_0^n$ By deforming f_n by a homotopy, we can suppose that $f_n(B^n) = y_0 \epsilon$ Y. Define f'_n to agree with f_n on $X^{(n)}$-int B^n, and define $f'_n: (B^n, \partial B^n) \to (Y, y_0)$ to represent $g \in \pi_n(Y, y_0)$. Clearly, "$f'_n - f_n$" $= \mu$. Hence, by the previous result $\tilde{\mathcal{O}}(f'_n) - \tilde{\mathcal{O}}(f_n) = \delta\mu$. ■

Let us consolidate our gains to this point.

Theorem 6.3. *Given* $f_n: X^{(n)} \cup A \to Y$ *with* Y *simply connected, there is a cohomology class* $\mathcal{O}(f_n) \in H^{n+1}(X, A; \pi_n(Y))$ *constructed from the cocycle* $\tilde{\mathcal{O}}(f_n)$. *This class vanishes if and only if* $f_n|X^{(n-1)} \cup A$ *can be extended to a map* $f: X^{(n+1)} \cup A \to Y$.

Let f, g: X → Y *be given and let* H: $(X^{(n)} \cup A) \times I \to Y$ *be a homotopy between the restrictions of* f *and* g *to* $X^{(n)} \cup A$. *The obstruction to extending the homotopy over* $(X^{(n+1)} \cup A) \times I$ *lies in*

$$H^{n+1}(X \times I, (X \times (\{0\} \cup \{1\})) \cup A \times I); \pi_n(Y)).$$

By the suspension isomorphism, this group is

$$H^n(X, A; \pi_n(Y)).$$

Thus, the obstructions to constructing a homotopy between two maps f: X → Y *and* g: X → Y, *given a fixed homotopy on A, lie in* $H^n(X, A; \pi_n(Y))$.

6.3 Further Properties

As long as $H^i(X, \pi_i(Y))$ is 0, we have no obstructions to finding a homotopy between f and g both restricted to $X^{(i)}$.

Suppose that $H^n(X, \pi_n(Y))$ is the first nonzero group. It is an exercise (using the theorem one step at a time over the skeleta $X^{(i)}$, $i < n$) to show that, given f and g : X → Y, the first obstruction to finding a homotopy between them, $\mathcal{O} \in H^n(X, \pi_n(Y))$, is well defined; i.e., does not depend on the step-by-step homotopy constructed from $f|X^{(n-1)}$ to $g|X^{(n-1)}$ this exhibits a universal phenomenon: **The first obstruction lying in a nonzero group is always well defined.** The higher obstructions are not defined until we make choices and then depend on these choices.

As an unproven example, take $\mathbb{C}P^2 \xrightarrow{\varphi} S^2$ where $\varphi|\mathbb{C}P^1 = *$ (= base point in S^2) and φ on the 4-cell is given by a nonzero element in $\pi_4(S^2)$; namely, $S^4 \xrightarrow{\Sigma\eta} S^3 \xrightarrow{\eta} S^2$ where $\Sigma\eta$ is the suspension of the Hopf map η. (We will do exercises of this type at a later time.) Take the point homotopy of $\varphi|\mathbb{C}P^1$ to $*$. Then the obstruction to extending this to a homotopy of φ to $*$ is in $H^4(\mathbb{C}P^2, \pi_4(S^2)) \cong \pi_4(S^2)$ and is the nonzero element ($\pi_4(S^2) \cong \mathbb{Z}_2$). But φ is homotopic to $*$. (Notice that $H^4(\mathbb{C}P^2, \pi_4(S^2))$ is not the first nonzero obstruction group since $H^2(\mathbb{C}P^2, \pi_2(S^2))$ is nonzero.)

Thus, obstruction theory is not the be all and end all of homotopy theory. It was once described by Sullivan as much like being in a labyrinth with a weak miner's light attached to your forehead and being forced always to move forward. The light enables you to see if you may take your next step but is not strong enough to tell you which fork to take when you must make a decision. Also there is no guarantee that if you choose one path that eventually is blocked, then all of them are blocked. On the other hand, if you were a miner would you care to be without your light? Likewise, no topologist would forsake obstruction theory.

6.4 Obstruction to the Existence of a Section of a Fibration

There is another type of obstruction theory which we shall need later on. This is for the problem of constructing a section of a fibration σ. Let $p: E \to B$ be a fibration with B path connected and with $F = p^{-1}(b_0)$. A section $\sigma: B \to E$ is a map for which $p \circ \sigma = Id_B$. We assume that B is a CW complex, hat $\pi_1(B)$ acts trivially on F, and that $\pi_1(F) = 0$. The obstructions to constructing a section lie in $H^i(B; \pi_{i-1}(F))$. Given two sections the obstructions to constructing a homotopy of sections connecting them lie in $H^i(B; \pi_{i-1}(F))$.

Let us give the definition of the obstruction cocycle. Suppose $\sigma: B^{(n-1)} \to E$ is a section over the (n − 1)-skeleton. For each n-cell of B, we have a trivialization of $p^{-1}(e^n) \to e^n$ as $e^n \times F$, as we saw in Chap. 5. To construct such a trivialization, one must connect e^n to b by a path in B. Since $\pi_1(B)$ acts trivially on the fiber, the identification is well defined, up to homotopy, independently of the path chosen. A section σ on $B^{(n-1)}$ induces a section of $p^{-1}(e^n) \to e^n$, denoted $\sigma|_{e^n}$. Using the above product structure, this gives $\tilde{\sigma}: e^n \to e^n \times F$. Projecting onto the second factor yields an element in $\pi_{n-1}(F)$. The obstruction cocycle $\tilde{\mathcal{O}}_n(\sigma): C_n(B) \to \pi_{n-1}(F)$ assigns to $[e^n]$, the resulting homotopy element.

Arguments similar to the ones above show that $\tilde{\mathcal{O}}_n(\sigma)$ is a cocycle and that one may vary $\tilde{\mathcal{O}}_n(\sigma)$ by an arbitrary coboundary by changing σ on $B^{(n-1)} - B^{(n-2)}$. Thus, the class $\mathcal{O}_n(\sigma) \in H^n(B; \pi_{n-1}(F))$ is the obstruction to extending $\sigma|B^{(n-2)}$ over $B^{(n)}$.

If (B, A) is a CW-pair and we are given a section over $B^{(n)} \cup A$, then the obstruction to extending its restriction to $B^{(n-1)} \cup A$ over $B^{n+1} \cup A$ lies in $H^{n+1}(B, A; \pi_n(F))$.

6.5 Examples

Let us consider some examples of these obstruction classes.

Example A. The Euler class: If $E^n \to B$ is an n-dimensional vector bundle, then a nowhere zero section of E^n is the same as a section of the associated sphere bundle $S^{n-1}(E)$. The first obstruction to finding a section is in $H^n(B, \pi_{n-1}(S^{n-1})) \cong H^n(B, \mathbb{Z})$. It is called the Euler class of E^n and is an *unstable characteristic class* of the vector bundle, unstable in the sense that taking the connected sum of the bundle with a trivial bundle kills the class. Some of the exercises deal with it in more detail.

Notice that this shows that if dim B < n, then $E^n \to B$ always has a nonzero section, and thus, as a vector bundle $E^n \cong E^{n-1} \oplus \epsilon^1$, where ϵ^1 is the trivial line bundle.

Example B. Eilenberg–MacLane spaces K(π, n): Given $n \geq 2$ and π an abelian group, we will show that up to homotopy equivalence, there is exactly one CW complex $K(\pi, n)$ such that

$$\pi_i(K(\pi, n)) = \begin{cases} \pi & \text{if } i = n, \\ 0 & \text{otherwise.} \end{cases}$$

Suppose that π is finitely presented

$$0 \to F(r_1, \ldots, r_l) \to F(\alpha_1, \ldots \alpha_k) \to \pi \to 0$$

where $F(, \ldots,)$ is the free abelian group generated by the symbols inside the parentheses. Form the space $(\vee_{i=1}^{k} S^{n}_{\alpha_i})$. Its nth homotopy group is $F(\alpha_1, \ldots, \alpha_k)$ (here we assume $n > 1$). Thus, the elements r_i are represented by maps $\varphi_{r_j}: S^n \to \vee_{i=1}^{k} S^{n}_{\alpha_i}$. Form the $(n+1)$-complex $\left(\vee_{i=1}^{k} S^{n}_{\alpha_i}\right) \cup \left(\cup_{j=1}^{l} e_{r_j}^{n+1)}\right)$ where the attaching map for $e_{r_j}^{n+1}$ is φ_{r_j}. Call this CW complex $X^{(n+1)}$. Clearly $\pi_n(X^{(n+1)}) = \pi$ and $\pi_i(X^{(n+1)}) = 0$ if $i < n$. Choose a generating set $\{\gamma_s\}_{s \in S}$ for $\pi_{n+1}(X^{(n+1)})$, and use these elements to attach $(n+2)$ -cells. Call the result $X^{(n+2)}$ One sees that

$$\pi_i(X^{(n+2)}) = \begin{cases} 0 & \text{if } i < n+2 \text{ and } i \neq n, \\ \pi & \text{if } i = n. \end{cases}$$

Continuing in this manner, we construct

$$X^{(n)} \subset X^{(n+1)} \subset X^{(n+2)} \subset \ldots$$

so that $\pi_i(X^{(n+k)}) = 0$ for $i < n + k$ and $i \neq n$, whereas $\pi_n(X^{(n+k)}) = \pi$. Let X be the union of the $X^{(n+k)}$. It has the correct homotopy groups.

There is a similar construction even if π is not finitely presented.

Now let us show that if Y is a CW complex with the same homotopy groups as X, then Y is homotopy equivalent to X. To establish this, we construct $f: X \to Y$ which induces an isomorphism on the homotopy groups. Begin with a map $f_n: \vee_{i=l}^{k} S^{n}_{\alpha_i} \to Y$ which induces the composition:

$$\pi_n(\vee S^{n}_{\alpha_i}) = F(\alpha_1, \ldots, \alpha_k) \to \pi = \pi_n(Y).$$

This map extends over $X^{(n+1)}$ since

$$\varphi_{r_j}: S^n \longrightarrow \vee_{i=1}^{k} S^{n}_{\alpha_i}$$

represents an element in $\pi_n(\vee S^{n}_{\alpha_i})$ which is trivial in $\pi = \pi_n(Y)$. Any extension $f_{n+1}: X^{(n+1)} \to Y$ induces the identity

$$\pi_n(X^{(n+1)}) = \pi \xrightarrow{\text{id}} \pi = \pi_n(Y).$$

The obstructions to extending f_{n+1} over all of X lie in $H^*(X; \pi_{*-1}(Y))$ for $* \geq n+2$. Hence, all obstructions vanish and f_{n+1} extends to $f: X \to Y$. Since $f_*: \pi_n(X) \to \pi_n(Y)$ is an isomorphism, by the Whitehead theorem, f is a homotopy equivalence.

Remark. $K(\mathbb{Z}, 1) = S^1$; $K(\mathbb{Z}/p\mathbb{Z}, 1), 1)$ is the infinite lens space, that is to say the quotient of the unit sphere in $\mathbb{C}^\infty$ by the action of the pth-roots of unity; and $K(\mathbb{Z}, 2) = \mathbb{C}P^\infty$, but outside of these the $K(\pi, n)$ are usually not spaces encountered directly in geometry. In general, the price for having the π_i so simple is that the homology $H_i(K(\pi, n), \mathbb{Z})$ is now very complicated (this seems quite plausible from the construction, since we are adding cells in all dimensions according (roughly) to the pattern of $\pi_k(S^n)(k > n)$).

Using the $K(\pi, n)$, we can construct a space with preassigned homotopy groups $\{\pi_i(X) = \pi_i\}$. Simply take $X = \prod_i K(\pi_i, n_i))$. Later we will show that any simply connected CW complex is homotopy equivalent to an iterated fibration of the $K(\pi, n)$.

For the moment we want to record one important fact about obstruction classes: the first possible nonzero class is well defined and natural. We establish this in the two contexts of extending a map and a section.

Theorem 6.4. *Let* (X, A) *be a CW-pair, and let* $f: A \to Y$ *be given. Suppose* $H^i(X, A; \pi_{i-1}(Y)) = 0$ *and* $H^{i-1}(X, A; \pi_{i-1}(Y)) = 0$ *for* $i \leq n$. *Also suppose* $\pi_1(Y) = 0$. *The first obstruction* $\mathcal{O} \in H^{n+1}(X, A; \pi_n(Y))$ *to extending* f *over* X *is well defined. It is natural with respect to maps* $\varphi: (X', A') \to X, A)$.

Proof. As we extend f over $X^{(i)} \cup A$, all obstructions lie in the zero group until we get to $X^{(n)} \cup A$. Let f_n and f'_n be two extensions over $X^{(n)} \cup A$. Since $H^i(X, A; \pi_i(Y)) = 0$ for $i \leq n-1$, $f'_n|(X^{(n)} \cup A)$ is homotopic relative to A to $f_n|(X^{(n)} \cup A)$. Thus, by 5.2, $\mathcal{O}(f_n) = \mathcal{O}(f'_n)$ in $H^{n+1}(X), A; \pi_n(Y)$. This proves that the primary obstruction is well defined. ■

Now suppose given $\varphi: (X', A') \to (X, A)$ where $H^*(X', A'; \pi_{*-1}(Y))$ and $H^{*-1}(X', A'\pi * -1(Y))$ are equal to zero for $* \leq n$. Then the first obstruction to extending $f \circ \varphi : A' \to Y$ lies in $H^{n+1}(X', A'; \pi_n(Y))$. As above we see that it is well defined. Naturality means that $\varphi^*(\mathcal{O}(f)) = \mathcal{O}(f \circ \varphi)$ in $H^{n+1}(X', A'; \pi_n(Y))$. To establish this, deform φ to a cellular map, φ', relative to A'. Let $f_n: X^{(n)} \cup A \to Y$ be an extension of $f: A \to Y$. Then $f_n \circ \varphi': X'^{(n)} \cup A \to Y$ is an extension of $f \cdot \varphi$ on A'. We claim that the cocycle obstruction for $f_n \cdot \varphi'$ is the pull back via φ' of the cocycle obstruction for f_n.

If $\varphi'_*(\sigma_1) = \Sigma_i \alpha_{1i} \tau_i$ in $C_{n+1}(X, A)$, then $\varphi'_*(\partial\sigma_1) = \Sigma_i \alpha_{1i} \partial\tau_i$ in $\pi_n(X^{(n)} \cup A)$. Thus,

$$\begin{aligned} < \mathcal{O}(f_n \cdot \varphi'), \sigma_i > &= f_n \circ \varphi'(\partial\sigma_i) \\ &= f_n(\Sigma_i \alpha_{1i} \partial\tau_i) \\ &= \Sigma_i \alpha_{1i} f_n(\partial\tau_i) \end{aligned}$$

$$= \Sigma_i \alpha_{1i} < \mathcal{O}(f_n), \tau_i >$$
$$= < \mathcal{O}(f_n), \Sigma_i \alpha_{1i} \tau_i >$$
$$= < \mathcal{O}(f_n), \varphi'_*(\sigma_1) >$$
$$= < (\varphi')^* \mathcal{O}(f_n), \sigma_1 > .$$ ■

Other Examples

C. If $S^n \xrightarrow{f} Y$, then the obstruction to f being homotopic to zero is in $H^n(S^n; \pi_n(Y)) = \tau_n(Y)$. It is, of course, the homotopy class of f in $\pi_n(Y)$.
D. Let $g: (\mathbb{C}P^2)^{(3)} \to S^2$ be the identity map. The obstruction to extending g over $\mathbb{C}P^2$ is in $H^4(\mathbb{C}P^2; \pi_3(S^2)) = \pi_3(S^2)$. It is the Hopf map; i.e., the attaching map for the 4-cell of $\mathbb{C}P^2$. This is true in general. If $X' = X \cup_f e^n$, then the obstruction to extending Id: $X \to X$ over X' is $[f] \in \pi_{n-1}(X)$.
E. Given $f: S^k \to Y$ and $g: S^\ell \to Y$, form $f \vee g: S^k \vee S^\ell \to Y$. The only obstruction to extending $f \vee g$ to a map $S^k \times S^\ell \to Y$ is an element in $H^{k+\ell}(S^k \times S^\ell, S^k \vee S^\ell; \pi_{k+\ell-1}(Y) \cong \pi_{k+\ell-1}(Y)$. Since this obstruction is primary, it is well defined. The obstruction, denoted [f, g], is the *Whitehead product* of f and g.

There is an analogous theorem for lifting maps in a fibration.

Theorem 6.5. *Let* (X, A) *be a relative CW complex and let* $\pi: E \to X$ *be a fibration with fiber* F. *Suppose that* $\pi_1(X)$ *acts trivially on* F *and that* $\sigma: A \to E|A$ *is a section of* π *over* A. *If* $H^i(X, A; \pi_{i-1}(F)) = 0$ *and* $H^i(X, A; \pi_i(F)) = 0$ *for* $i \leq n$, *then the first obstruction to extending* σ *to a section defined on all of* X *lies in* $H^{n+1}(X, A; \pi_n(F))$. *It is well defined and natural.*

The proof is analogous to that of Theorem 6.4 and is left to the reader.

Chapter 7
Eilenberg–MacLane Spaces, Cohomology, and Principal Fibrations

7.1 Relation of Cohomology and Eilenberg–MacLane Spaces

In the last section, we saw that there was a natural transformation

$$[(X, A), (K(\pi, n), *)] \to H^n(X, A; \pi)$$

which assigns to any map $f: (X, A) \to (K(\pi, n), *)$, the primary obstruction to deforming f to a constant map relative to A. (Here, $*$ is the base point of $K(\pi, n)$.) Actually, this should be viewed as an extension problem for the map $f \cup c \cup c$: $X \times \{0\} \cup A \times I \cup X \times \{1\} \to K(\pi, n)$ (where c denotes the constant map to the base point). The primary obstruction is well defined and lies in

$$H^{n+1}(X \times I, X \times \{0\} \cup A \times I \cup X \times \{1\}\,; \pi) \cong H^n(X, A; \pi).$$

By the Hurewicz theorem, $H_n(K(\pi, n); \mathbb{Z}) = \pi$ and $H_{n-1}(K(\pi, n); \mathbb{Z}) = 0$. Thus, by the universal coefficient theorem, $H^n(K(\pi, n); \pi) = \mathrm{Hom}(\pi, \pi)$. (Here, we are viewing $K(\pi, n)$ as a space together with an identification of $\pi_n(K(\pi, n))$ with π.) Let $\iota \in H^n(K(\pi, n); \pi)$ be the class corresponding to the identity homomorphism $\pi \to \pi$. If $f: (X, A) \to (K(\pi, n), *)$, then we have $f^*(\iota) \in H^n(X, A; \pi)$. This defines a function

$$[(X, A), (K(\pi, n), *)] \to H^n(X, A; \pi)$$
$$f \mapsto f^*(\iota).$$

Theorem 7.1. *The class* $f^*\iota$ *is the primary obstruction to deforming* f *to the constant map relative to* A. *The association* $[f] \mapsto f^*\iota$ *is a bijection for all CW-pairs* (X, A).

P. Griffiths and J. Morgan, *Rational Homotopy Theory and Differential Forms*, Progress in Mathematics 16, DOI 10.1007/978-1-4614-8468-4_7,
© Springer Science+Business Media New York 2013

Proof. First we show that $f^*\iota$ is the first obstruction to deforming f to a constant map relative to A. By naturality it suffices to show that the obstruction for the identity map $K(\pi, n) \to K(\pi, n)$ is ι. This is immediate from the definitions.

Since $K(\pi, n)$ has only one nonzero homotopy group, if $f^*\iota = 0$, then the one and only obstruction to deforming f to a point, relative to A, vanishes. It follows that in that case f is homotopic, rel A, to a constant map. This shows that $\iota^{-1}(0)$ is exactly the trivial map. This does not yet suffice to prove that ι is 1–1 because we have not defined a group structure on $[(X, A), (K(\pi, n), *)]$, cf. below.

We show that ι is onto. Let $\alpha \in H^n(X, A; \pi)$ be a class, and let $\tilde{\alpha} : C_n(X, A) \to \pi$ be a cocycle representative for α. Define $f_{\tilde{\alpha}}|X^{(n-1)} \cup A$ to be the constant map. For each n-call e^n define $(f_{\tilde{\alpha}}|_{e^n})\colon e^n \to K(\pi, n)$ to be a map representing $\tilde{\alpha}(e^n) \in \pi = \pi_n(K(\pi, n))$. This gives $f_{\tilde{\alpha}}|X^{(n)} \cup A$. Since $\tilde{\alpha}$ is a cocycle, if we compose $f_{\tilde{\alpha}}|X^{(n)} \cup A$ with the attaching map for an $(n + 1)$-cell, the result is homotopically trivial. Thus, $f_{\tilde{\alpha}}|X^{(n)} \cup A$ extends over $X^{(n+1)} \cup A$. All higher obstructions to extending $f_{\tilde{\alpha}}|X^{(n+1)} \cup A$ vanish. Let $f_{\tilde{\alpha}}\colon (X, A) \to K(\pi, n)$ be an extension of $f_{\tilde{\alpha}}|X^{(n)} \cup A$. Clearly, the relative cocycle which measures the primary obstruction to a homotopy from $f_{\tilde{\alpha}}$ to the constant map is exactly $\tilde{\alpha} \in \mathrm{Hom}(C_n(X, A), \pi)$. Thus, $f^*_{\tilde{\alpha}}(\iota) = \alpha$.

To see that ι is 1–1, we assume that f and g map (X, A) to $(K(\pi, n), *)$ and $f^*\iota = g^*\iota$. The primary (and only) obstruction to a homotopy, relative to A, from f to g is in $H^n(X, A; \pi)$. It is easily identified with $f^*(\iota) - g^*(\iota) = 0$. Thus, f is homotopic to g relative to A. ■

The Additive Structure: There is a map

$$K(\pi, n) \times K(\pi, n)| \xrightarrow{\mu} K(\pi, n)$$

so that

$$\mu^*\iota = \iota \otimes 1 + 1 \otimes \iota \in H^n(K(\pi, n) \times K(\pi, n); \pi) \cong$$

$$\cong [H^n(K(\pi, n); \pi) \otimes H^0(K(\pi, n); \mathbb{Z})] \oplus [H^0(K(\pi, n); \mathbb{Z}) \otimes H^n(K(\pi, n);\ \pi)]$$

It is well defined up to homotopy. It acts as a commutative, associative group multiplication up to homotopy. Thus, $K(\pi, n)$ is a *homotopy commutative, associative H-space*. This map μ induces an abelian group structure on $[(X, A); (K(\pi, n), *)]$. with this group structure ι becomes a group isomorphism. (Details of proof left as exercises.)

7.2 Principal $K(\pi, n)$-Fibrations

A map $p\colon E \to B$ is said to be a $K(\pi, n)$ fibration if it satisfies the homotopy lifting property and if all fibers, $p^{-1}(b)$, are spaces of type $K(\pi, n)$. If B is path connected, then $p\colon E \to B$ is a $K(\pi, n)$-fibration provided that $p^{-1}(b)$ is a space of type $K(\pi, n)$ for any $b \in B$.

A $K(\pi, n)$ fibration is said to be *principal* if the action of the fundamental group of the base on the fiber is trivial up to homotopy. For each loop γ in the base B based at b there is a self-homotopy equivalence $\gamma_* : p^{-1}(b) \to p^{-1}(b)$ well defined up to homotopy. The fibration is principal if all these self-equivalences are homotopic to the identity.

Lemma 7.2. *Let* (X, A) *be a CW pair, let* $p: E \to X$ *be a principal* $K(\pi, n)$*-fibration, and let* $\sigma: A \to E$ *be a section of* E *over* A. *There is a unique obstruction* $\mathcal{O}(p, \sigma) \in H^{n+1}(A; \pi)$ *to extending* σ *over all of* X. *Given any class* $\mathcal{O} \in H^{n+1}(X, A; \pi)$ *is realized as the obstruction* $\mathcal{O}(p, \sigma)$ *for some principal fibration* $p: E \to X$ *and for some section* $\sigma: A \to E$.

Proof. Since $\pi_i(\text{fiber}) = 0$ for $i < n$, according to Theorem 6.3, the first obstruction to extending the section lies in $H^{n+1}(X, A; \pi)$. It is well defined and natural. Since all the higher homotopy groups of the fiber vanish, $\mathcal{O}(p, \sigma)$ is the unique obstruction to extending the section over all of X. Given a class $\mathcal{O} \in H^{n+1}(X, A; \pi)$, there is a map

$$f_{\mathcal{O}}: (X, A) \to (K(\pi, n+1), *)$$

so that $f_{\mathcal{O}}^*(\iota) = \mathcal{O}$. Over $K(\pi, n+1)$ we have the principal fibration

$$\begin{array}{ccc} K(\pi, n) = \Omega K(\pi, n+1) & \longrightarrow & \mathcal{P}(K(\pi, n+1)) \\ & & \downarrow \\ & & K(\pi, n+1) \end{array}$$

Since $K(\pi, n+1)$ is simply connected, this fibration is principal. If we pull back the fibration by $f_{\mathcal{O}}$, then we get a principal $K(\pi, n)$ fibration over X. Any section over $*$ pulls back to a section over A. The obstruction to extending the section over $* \in K(\pi, n+1)$ to one over all of $K(\pi, n+1)$ is $\iota \in H^{n+1}(K(\pi, n+1), \pi)$. By naturality the obstruction to extending the induced section over A to one over all of X is $f_{\mathcal{O}}^*(\iota) = \mathcal{O}$. This proves all classes arise as obstructions. ■

Suppose that $E \xrightarrow{f} B$ and $E' \xrightarrow{f'} B$ are principal $K(\pi, n)$-fibrations. In particular, the fibers are identified with $K(\pi, n)$ uniquely up to homotopy. They are said to be *equivalent* if there is a map $\Phi: E \to E'$ with $f' \circ \Phi = f$ and with Φ inducing a map on the fibers which is compatible up to homotopy with the identification of the fibers with $K(\pi, n)$, meaning that after these identifications, the map given by the restriction of Φ induces the identity on the nth homotopy group of the fibers. More generally given a pair (B, A), two principal $K(\pi, n)$-fibrations $E \to B$ and $E' \to B$ with sections σ and σ' over A are *equivalent* if there is a homotopy equivalence $\Phi: E \to E'$ computing with the projections to B with the property that $\Phi \circ \sigma$ is homotopy to σ' as sections of $E'|_A$.

There is also a relative version of these results summarized in the following commutative diagram.

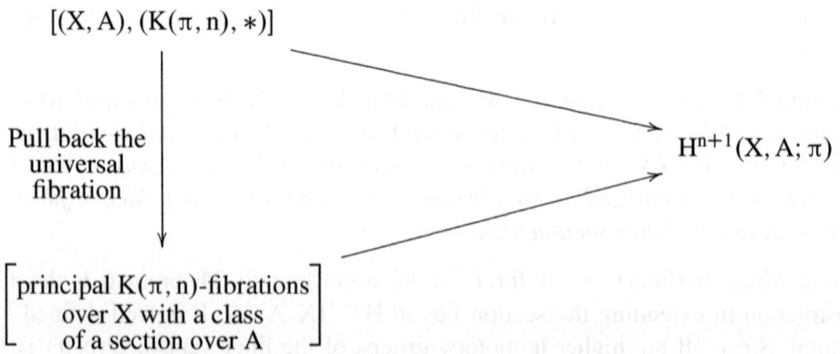

where diagonal arrows are the obstruction to a relative homotopy to the constant map and the obstruction to the extension of the section over A, respectively.

All maps are bijections. To establish the relative version, use the fact that $H^{n+1}(X, A; \pi) = H^{n+1}(X/\{A\}; \pi)$, where $X/\{A\}$ is the CW complex obtained from X by collapsing A to a point. Thus, elements in $H^{n+1}(X, A; \pi)$ correspond to principal $K(\pi, n)$-fibrations over $X/\{A\}$. These fibrations have a section over the point $\{A\}$, and this section is unique up to homotopy. Pulling back via the natural map $(X, A) \to X/\{A\}, \{A\})$ gives the result.

We say that a map $f\colon E \to B$ is homotopy equivalent to a principal $K(\pi, n)$-fibration if there is a principal $K(\pi, n)$-fibration $\tilde{E} \to B$ and a homotopy equivalence $\varphi\colon E \to \tilde{E}$ that commutes with the maps to B. It is an easy exercise using obstruction theory to show the following:

Proposition 7.3. *Let* B *and* E *be simply connected CW complexes. A map* $f\colon E \to B$ *is homotopy equivalent to a principal* $K(\pi, n)$*-fibration if and only if the following two conditions hold:*

(i) $H_*(B, E) = 0$ *for* $* \neq n + 1$ *and*
(ii) $H_{n+1}(B, E) = \pi$.

Suppose that B is a CW complex and that $\tilde{E} \to B$ is a principal $K(\pi, n)$-fibration. Then $\tilde{E}$ is homotopy equivalent to a CW complex. This is proved by induction over the cells of B; details are left to the reader.

Proposition 7.4. *Let* $p\colon \tilde{E} \to B$ *be a principal* $K(\pi, n)$*-fibration over a simplicial complex. Then there is a simplicial complex* E *and a homotopy equivalence* $\varphi\colon E \to \tilde{E}$ *so that the composition* $p \circ \varphi\colon E \to B$ *is a simplicial map.*

Proof. Since $\tilde{E}$ is homotopy equivalent to a CW complex, there is a simplicial complex E and a homotopy equivalence $\varphi_0\colon E_0 \to \tilde{E}$. At the expense of subdividing E, we can assume that with the map $p \circ \varphi_0\colon E \to B$ is homotopic to a simplicial map

$\psi: E \to B$. Applying the homotopy lifting property, we can deform φ_0 to a map $\varphi: E \to \tilde{E}$ with $p \circ \varphi = \psi$. ■

Definition 7.5. We say that $\psi: E \to B$ is a *simplicial model* for the principal $K(\pi, n)$-fibration over B.

Chapter 8
Postnikov Towers and Rational Homotopy Theory

Fibrations and CW structures should be viewed as dual, and a Postnikov tower for a space is a decomposition dual to a cell decomposition. In the Postnikov tower description of a space, the atoms of the space (which we think of as a molecule) are $K(\pi, n)$s. (These are atomic from the point of view of homotopy groups. The spheres are atomic from the point of view of homology groups.) The geometric configuration of the atoms in the molecule is exactly the information contained in the "k-invariants" of the space. These tell us how the various $K(\pi, n)$s are twisted together. We shall prove that all simply connected CW complexes have such towers.

Given a simply connected space X, define $X_2 = K(\pi_2(X), 2)$, and define $f_2 \colon X \to X_2$ to be a map inducing the identity on π_2. Suppose inductively that we have a commutative diagram, up to homotopy,

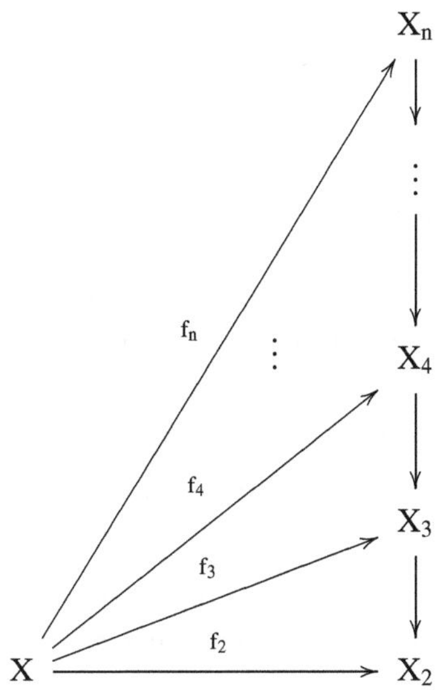

P. Griffiths and J. Morgan, *Rational Homotopy Theory and Differential Forms*, Progress in Mathematics 16, DOI 10.1007/978-1-4614-8468-4_8,

© Springer Science+Business Media New York 2013

such that for every $j \leq n$, we have

(1) $\pi_i(X_j) = 0$ for all $i > j$.
(2) The map $X_j \to X_{j-1}$ is a principal fibration with fiber $K(\pi_j(X), j)$ induced by some map $k^{j+1}: X_{j-1} \to K(\pi_j(X), j+1)$.
(3) $f_j: X \to X_j$ an isomorphism on π_i for all $i \leq j$.

Invoking the mapping cylinder construction allows us to assume that $f_n : X \to X_n$ is an inclusion. The relative homology $H_i(X_n, X)$ is zero for $i \leq n+1$. Furthermore,

$$H_{n+2}(X_n, X) \cong \pi_{n+2}(X_n, X) \xrightarrow{\sim} \pi_{n+1}(X).$$

By the universal coefficient theorem,

$$H^{n+2}(X_n, X; \pi_{n+1}(X)) = \mathrm{Hom}(\pi_{n+1}(X), \pi_{n+1}(X)).$$

Let $\tilde{k}^{n+1} \in H^{n+2}(X_n, X; \pi_{n+1}(X))$ be the class corresponding to the identity homomorphism. This determines a principal fibration $X_{n+1} \to X_n$ with a section over X. The section over X is equivalently and a lifting of f_{n+1} of f_n as indicated in the diagram below.

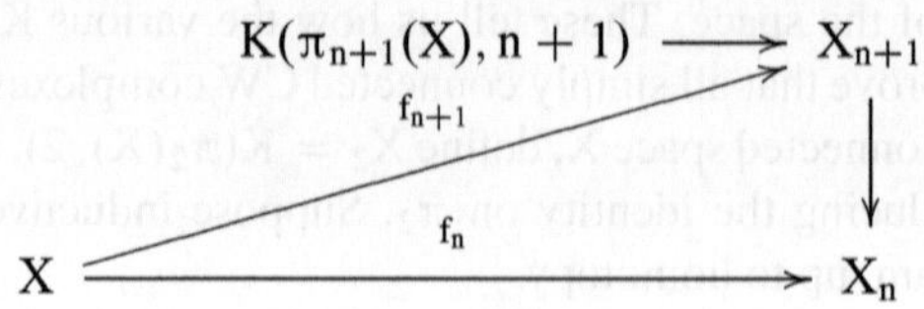

Lemma 8.1. *The map* $f_{n+1} : X \to X_{n+1}$ *meets conditions* (1), (2), (3) above.

Proof. Clearly (1) and (2) are satisfied. Also, $(f_{n+1})_*: \pi_i(X) \to \pi_i(X_{n+1})$ is an isomorphism for $i \leq n$. It remains to show that $(f_{n+1})_*: \pi_{n+1}(X) \to \pi_{n+1}(X_{n+1})$ is an isomorphism. We have a map of $(X_n, X) \to (X_n, X_{n+1})$. If this map induces an isomorphism $\pi_{n+2}(X_n, X) \xrightarrow{\sim} \pi_{n+2}(X_n, X_{n+1})$, then it follows that $(f_{n+1})_*$ is an isomorphism on π_{n+1}. Under the identification of $\pi_{n+2}(X_n, X_{n+1})$ with $\pi_{n+1}(K(\pi_{n+1}(X), n+1)) = \pi_{n+1}(X)$, the map becomes evaluation of the cohomology class $\tilde{k}^{n+2}$. By definition this map is an isomorphism. ■

Remarks. (1) For each n let X'_n be the CW complex obtained from X by inductively attaching cells of dimension $\geq n+2$ so as to kill all homotopy groups in dimensions $\geq n+1$, so that $\pi_i(X'_n) = 0$ for $i > n$. The inclusion $X \subset X_n$ induces an isomorphism on π_i for $i \leq n$.

If $f_n: X \to X_n$ is the nth stage of a Postnikov tower for X, then a simple application of obstruction theory shows that f_n extends to a map $f'_n: X'_n \to X_n$. This map induces an isomorphism on all homotopy groups, showing that there is a homotopy equivalence from X'_n to the nth stage of the Postnikov tower of X, a homotopy equivalence that is compatible up to homotopy with the natural maps from X to these spaces.

If $g_n : X \to Y_n$ is the nth stage of another Postnikov tower for X, then we have a commutative diagram

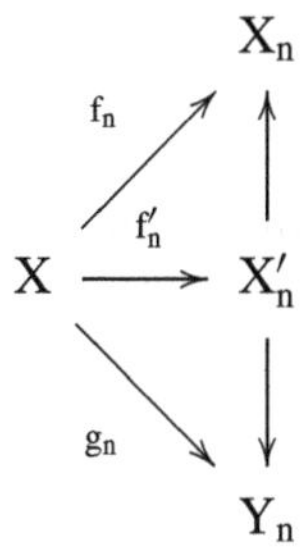

with both vertical arrows being homotopy equivalences. Under the resulting identifications of $H^{n+2}(X_n, X; \pi_{n+1}(X))$ with $H^{n+2}(Y_n, X; \pi_{n+1}(X))$ the k-invariants for the $(n+1)$st-stages of the two towers correspond, as is easy to see. It is in this sense that the Postnikov tower is unique in the homotopy category.

(2) If we form $\varprojlim\{X_i\}$, defined as the subspace of $\prod_{i=2}^{\infty} X_i$ consisting of all compatible sequences, then the maps $\{f_n : X \to X_n\}$ determine $\varprojlim f_n : X \to \varprojlim\{X_i\}$. This map induces an isomorphism on all homotopy groups. To prove this, one shows that $\pi_j(\varprojlim X_n) = \varprojlim \pi_j(X_n) = \pi_j(X_N)$ for $N \geq j$. It is not true in general for inverse systems that taking homotopy groups commutes with taking inverse limits. It is, however, true for this inverse system since $\pi_{j+1}(X_N) \to \pi_{j+1}(X_{N-1})$ is onto for all N.

(3) If Y is a CW complex and $\varphi : Y \to \varprojlim X_n$ induces an isomorphism on all homotopy groups, then the obstructions to lifting, up to homotopy, $\varprojlim f_n : X \to \varprojlim X_n$ to Y lie in $H^*(X; \pi_*(\varprojlim X_n, Y)) = 0$. Thus, there is a map $\psi : X \to Y$ such that $\varphi \circ \psi$ is homotopic to $\varprojlim f_n$. In particular ψ induces an isomorphism on all homotopy groups, and hence, ψ is a homotopy equivalence.

This shows how to recover X, up to homotopy equivalence, from a Postnikov tower: It is the unique CW complex, up to homotopy equivalence, which maps to $\varprojlim X_n$ inducing an isomorphism on the homotopy groups.

(4) Suppose X is a CW complex of dimension n. Once we have built the Postnikov tower for X through dimension n, the process of finishing the tower is formal (if complicated). Namely, $f_n : X \to X_n$ induces an isomorphism on H_i for $i \leq n$ and is surjective on H_{n+1}. Thus, $H_{n+1}(X_n) = 0$, but $H_{n+2}(X_n)$ may not be zero.

Suppose it is π. Then the $(n + 1)$st homotopy group of X is π and the k-invariant $k^{n+2} \in H^{n+2}(X_n, \pi)$ is the identity homomorphism. By the Serre spectral sequence for

$$\begin{array}{ccc} K(\pi_{n+1}(X), n+1) & \longrightarrow & X_{n+1} \\ & & \downarrow \\ & & X_n \end{array}$$

one has $H_{n+2}(X_{n+1}) = 0$. We continue in this fashion:

$$H_{n+k+1}(X_{n+k}) = 0, \quad H_{n+k+2}(X_{n+k}) = \pi_{n+k+1}(X),$$

and the k-invariant is the identity map:

$$H_{n+k+2}(X_{n+k}) \longrightarrow \pi_{n+k+1}(X).$$

This ability to complete the Postnikov tower for X once we have X_n without making further reference to the space X is referred to. By saying that, after X_n, the process becomes *formal*. Thus, all simply connected spaces X of homological dimension n are formal after X_n.

(5) Let us consider the case $X = S^2$. Then $X_2 = K(\mathbb{Z}, 2) = \mathbb{C}P^\infty$, $H_3(\mathbb{C}P^\infty) = 0$ and $H_4(\mathbb{C}P^\infty) = \mathbb{Z}$. Hence, $\pi_3(S^2) = \mathbb{Z}$. If we form

$$\begin{array}{ccc} K(\mathbb{Z}, 3) & \longrightarrow & X_3 \\ & & \downarrow \\ & & \mathbb{C}P^\infty \end{array}$$

with k-invariant the identity $H_4(\mathbb{C}P^\infty) \to \mathbb{Z}$, then $H_4(X_3) = 0$ and $H_5(X_3) = \mathbb{Z}/2$. Hence, $\pi_4(S^2) = \mathbb{Z}/2$. Continuing in this way gives a (theoretical) algorithm for calculating all the higher homotopy groups of S^2. This calculation has never been done and seemingly is impossibly complicated to do. It is an amazing theorem of E. Curtis that $\pi_i(S^2) \neq 0$ for all $i \geq 2$.

According to the last section given a Postnikov tower, we can replace each of the fibrations by a simplicial model.

$$\begin{array}{ccc}
\downarrow & & \downarrow \\
X_n & \xleftarrow{\varphi_n} & B_n \\
\downarrow & & \downarrow \psi_n \\
X_{n-1} & \xleftarrow{\varphi_{n-1}} & B_{n-1} \\
\downarrow & & \downarrow \\
\cdot & & \cdot \\
\cdot & & \cdot \\
\downarrow & & \downarrow \psi_3 \\
X_2 & \xleftarrow{\varphi_2} & B_2.
\end{array}$$

Here the B_i are simplicial complexes, the maps ψ_i are simplicial maps, and the φ_i are homotopy equivalences. We call the tower of simplicial complexes a *simplicial model* for the Postnikov tower.

8.1 Rational Homotopy Theory for Simply Connected Spaces

We begin with a little of the theory of $\mathbb{Q}$ and $\mathbb{Q}$-vector spaces. Let A be an abelian group (usually infinitely generated). Then A may be given the structure of a $\mathbb{Q}$ vector space if and only if $A \cong A \otimes_{\mathbb{Z}} \mathbb{Z} \xrightarrow{\sim} A \otimes_{\mathbb{Z}} \mathbb{Q}$. (This is equivalent to the equation $\alpha x = \beta$ having a unique solution x for all $\alpha \in \mathbb{Z} - \{0\}$ and $\beta \in A$.)

If

$$0 \to A_1 \to A_2 \to A_3 \to 0$$

is a short exact sequence then so is

$$0 \to A_1 \otimes_{\mathbb{Z}} \mathbb{Q} \to A_2 \otimes_{\mathbb{Z}} \mathbb{Q} \to A_3 \otimes \mathbb{Q}_{\mathbb{Z}} \to 0.$$

Lemma 8.2. *(a) If* $0 \to A_1 \to A_2 \to A_3 \to 0$ *is a short exact sequence, then if two of the three terms are* $\mathbb{Q}$*-vector spaces, so is the third one.*

(b) If A *is an abelian group and has a composition series*

$$A = A_0 \supset A_1 \supset \ldots \supset A_n \supset 0$$

with successive quotients $\mathbb{Q}$*-vector spaces, then* A *is a* $\mathbb{Q}$*-vector space.*

(c) *For any space* X, *we have* $H_*(X;\mathbb{Q}) \cong H_*(X) \otimes_{\mathbb{Z}} \mathbb{Q}$ *and* $H^*(X;\mathbb{Q}) \simeq \mathrm{Hom}_{\mathbb{Z}}(H_*(X), \mathbb{Q}) \cong [H_*(X;\mathbb{Q})]^*$.

(d) *If* $\tilde{H}_*(X)$ *is a* $\mathbb{Q}$*-vector space, then* $\tilde{H}_*(X;G)$ *is a* $\mathbb{Q}$*-vector space for any abelian group* G.[1]

Proof.

(a) There is a commutative ladder of short exact sequences:

$$\begin{array}{ccccccccc} 0 & \longrightarrow & A_1 & \longrightarrow & A_2 & \longrightarrow & A_3 & \longrightarrow & 0 \\ & & \downarrow & & \downarrow & & \downarrow & & \\ 0 & \longrightarrow & A_1 \otimes \mathbb{Q} & \longrightarrow & A_2 \otimes \mathbb{Q} & \longrightarrow & A_3 \otimes \mathbb{Q} & \longrightarrow & 0 \end{array}$$

By hypothesis two of the three vertical maps are isomorphisms. Thus, by the 5 lemma, so is the third.

(b) Follows immediately by induction from (a).

(c) and (d) are left to the reader as exercises. ■

Corollary 8.3. $H_i(X;\mathbb{Q})$ *and* $H^i(X;\mathbb{Q})$ *are* $\mathbb{Q}$*-vector spaces.*

Definition. A $\mathbb{Q}$-space is a space, X, satisfying:

(1) X is homotopy equivalent to a CW complex (generally having ∞-many cells in each dimension).
(2) $\pi_1(X) = 0$.
(3) $\pi_*(X)$ is a $\mathbb{Q}$-vector space for all $* \geq 1$.

The following is an alternate definition. A space X satisfying (1) and (2) above is a $\mathbb{Q}$-space if in addition (3) $\tilde{H}_*(X, \mathbb{Z})$ is a $\mathbb{Q}$-vector space.

Theorem 8.4. *The two definitions of a* $\mathbb{Q}$*-space are equivalent.*

Proof. We will need the following lemma:

Lemma 8.5. (a) $\tilde{H}_*(K(\mathbb{Q}, n); \mathbb{Z})$ *is a* $\mathbb{Q}$*-vector space.*
(b) $H^*(K(\mathbb{Q}, 2n); \mathbb{Q})$ *is a* $\mathbb{Q}$*-polynomial algebra on one generator, that generator being of degree* 2n.
(c) $H^*(K(\mathbb{Q}, 2n+1); \mathbb{Q})$ *is a* $\mathbb{Q}$*-exterior algebra on one generator, that generator being of degree* $2n+1$.

Proof. We construct $K(\mathbb{Q}, 1) = X$ as an iterated mapping cylinder:

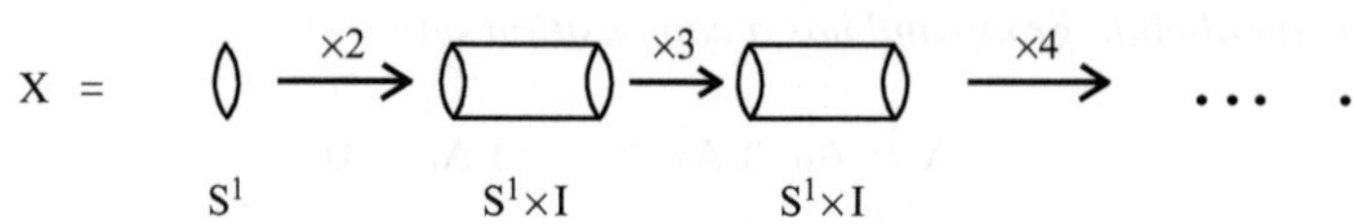

[1] $\tilde{H}_*$ = reduced homology groups.

Since homotopy commutes with direct limits,

$$\pi_1(X) = \varinjlim\{\mathbb{Z}, \times n\} = \mathbb{Q},$$

and

$$\pi_i(X) = \varinjlim\{0\} = 0 \quad \text{for} \quad i > 1.$$

Thus, X is a $K(\mathbb{Q}, 1)$. Since homology also commutes with direct limits

$$\tilde{H}_*(Z; \mathbb{Z}) = \begin{cases} \mathbb{Q} \text{ for } * = 1 \\ 0 \text{ for } * \neq 1. \end{cases}$$

Suppose inductively that we have shown that $\tilde{H}_*(K(\mathbb{Q}, n-1)); \mathbb{Z})$ is a $\mathbb{Q}$-vector space. Consider the Serre spectral sequence for

$$\begin{array}{ccc} K(\mathbb{Q}, n-1) & \longrightarrow & \mathcal{P}(K(\mathbb{Q}, n)) \\ & & \downarrow \\ & & K(\mathbb{Q}, n) \end{array}$$

with $\mathbb{Z}$-coefficients and with $\mathbb{Q}$-coefficients.

There is a comparison theorem for spectral sequences which says that, given two spectral sequences and a map between them inducing an isomorphism on $E_\infty^{*,*}$ for $(*, *) \neq (0, 0)$ and on $E_2^{0,q}$ for all $q > 0$, then it is an isomorphism on $E_2^{p,q}$ for all $(p, q) \neq (0, 0)$. In particular, it is an isomorphism on $E_2^{p,0}$ for $p > 0$. This result is an algebraic one proved by induction, cf. Exercise (42).

Applying it to the Serre spectral sequences for the above fibration with $\mathbb{Z}-$ and $\mathbb{Q}$-coefficients, we see that if $\tilde{H}_*(K(\mathbb{Q}, n-1); \mathbb{Z})$ is a $\mathbb{Q}$-vector space, then so is $\tilde{H}_*(K(\mathbb{Q}, n); \mathbb{Z})$.

Parts (b) and (c) are proved by the same inductive argument given in Chap. 5 to calculate $H^*(K(\mathbb{Z}, n); \mathbb{Q})$. ■

Corollary 8.6. $K(\mathbb{Z}, n) \to K(\mathbb{Q}, n)$ *induces an isomorphism on rational cohomology and thus on rational homology.*

Using Lemma 8.5 we will show that if $\pi_1(X) = 0$ and $\pi_i(X)$ is a $\mathbb{Q}$-vector space for all i, then $\tilde{H}_i(X)$ is a $\mathbb{Q}$-vector space for all i. Since $H_*(X) \cong H_*(X_n)$ for $* \leq n$, where X_n is the nth stage in the Postnikov system for X, it suffices to prove by induction that $\tilde{H}_*(X_n)$ is a $\mathbb{Q}$-vector space. Since $X_2 = K(\pi_2(X), 2)$ and $\pi_2(X)$ is a rational vector space, from Lemma 8.5, we see that $\tilde{H}_*(X_2)$ is a $\mathbb{Q}$-vector space. Suppose $\tilde{H}_*(X_{n-1})$ is a $\mathbb{Q}$-vector space and consider the Serre spectral sequence for

$$\begin{array}{ccc} K(\pi_n(X, n)) & \longrightarrow & X_n \\ & & \downarrow \\ & & X_{n-1} \end{array}$$

Now $E_2^{p,q}$ is a rational vector space for all $(p, q) \neq (0,0)$, and consequently $E_\infty^{p,q}$ is also a rational vector space for all $(p, q) \neq (0, 0)$. Thus, there are composition series for $\tilde{H}_i(X_n)$ with successive quotients $\mathbb{Q}$-vector spaces. This proves $\tilde{H}_i(X_n)$ is a $\mathbb{Q}$ -vector space for all i.

Conversely, suppose we have a space X with $\pi_1(X) = 0$ and $\tilde{H}_*(X)$ a $\mathbb{Q}$-vector space. We will show by induction that $\pi_i(X)$ is a $\mathbb{Q}$-vector space. Suppose we know that $\pi_*(X_{n-1})$ are $\mathbb{Q}$-vector spaces. From what we have just established, it follows that $\tilde{H}_*(X_{n-1})$ is a rational vector space. Hence, so is $\tilde{H}_*(X_{n-1}, X)$. In particular, $\pi_n(X) = H_{n+1}(X_{n-1}, X)$ is a rational vector space. Thus, all the homotopy groups of X_n are rational vector spaces. Continuing the induction, we establish that all the homotopy groups of X are rational vector spaces. ■

Theorem 8.7. *Let* X *and* $X_{(0)}$ *be simply connected CW complexes with* $X_{(0)}$ *a* $\mathbb{Q}$-*space and* $f: X \to X_{(0)}$. *The following three conditions are equivalent:*

(a) $f_*: \tilde{H}_*(X, \mathbb{Q}) \to \tilde{H}_*(X_{(0)}; \mathbb{Q}) = \tilde{H}_*(X_{(0)})$ *is an isomorphism*
(b) $f_\# : \pi_*(X) \otimes \mathbb{Q} \to \pi_*(X_{(0)}) \otimes \mathbb{Q} = \pi_*(X_{(0)})$ *is an isomorphism*
(c) f *is universal for maps of* X *into* $\mathbb{Q}$-*space; i.e., given* $g: X \to Y_{(0)}$ *with* $Y_{(0)}$ *a* $\mathbb{Q}$-*space, then* g *factors uniquely up to homotopy through* $X_{(0)}$

$$\begin{array}{ccc} X & \xrightarrow{\;g\;} & Y_{(0)} \\ {\scriptstyle f}\downarrow & \nearrow{\scriptstyle h} & \\ X_{(0)} & & \end{array}$$

Proof. We show that (a) implies (c). Given f satisfying (a) and $g: X \to Y_0$, then the obstructions to extending g over all of $X_{(0)}$ (considering f now as an inclusion) lie in $H^*(X_{(0)}, X; \pi_*(Y_{(0)}))$. Since $f_*: H_*(X) \to H_*(X_{(0)})$ is an isomorphism with $\mathbb{Q}$-coefficients and $\pi_*(Y_{(0)})$ is a $\mathbb{Q}$-vector space $H^{*+1}(X_{(0)}, X; \pi_*(Y_0)) = 0$. Thus, such a map h exists. The obstruction to any two maps being homotopic relative to g is in $H^*(X_{(0)}, X: \pi_*(Y_{(0)})) = 0$. This proves that (a) implies (c).

Now we show that (c) implies (a). Apply universality to $Y_{(0)} = K(\mathbb{Q}, n)$ and we see that there is a 1–1 correspondence

$$\begin{array}{ccc} [X, K(\mathbb{Q}, n)] & \xleftarrow[f^*]{\cong} & [X_{(0)}, K(\mathbb{Q}, n)] \\ \cong\uparrow & & \uparrow\cong \\ H^n(X, \mathbb{Q}) & \xleftarrow[f^*]{\cong} & H^n(X_{(0)}, \mathbb{Q}) \end{array}$$

and thus, f^* is an isomorphism on $\mathbb{Q}$-cohomology and thus on $\mathbb{Q}$-homology.

To prove (a) and (b) are equivalent, we need the following lemma:

Lemma 8.8. *If X is a simply connected CW complex, then* $\pi_*(X) \otimes \mathbb{Q} = 0$ *if and only if* $\tilde{H}_*(X; \mathbb{Q}) = 0$.

(This is a variant of the Hurewicz theorem modulo the Serre class of abelian groups A such that $A \otimes \mathbb{Q} = 0$.)

Proof.

Step 1: Suppose that $\pi = \mathbb{Z}/k\,\mathbb{Z}$. Then, $\tilde{H}_*(K(\pi, n); \mathbb{Q}) = 0$. To see this, notice that we have a fibration $K(\mathbb{Z}/k\mathbb{Z}, n) \to K(\mathbb{Z}, n+1) \xrightarrow{\cdot k} K(\mathbb{Z}, n+1)$. From the fact that $H^*(K(\mathbb{Z}, n+1); \mathbb{Q})$ is either a polynomial algebra or an exterior algebra, it follows that $(\cdot k)\colon H^*(K(\mathbb{Z}, n+1); \mathbb{Q}) \to H^*(K(\mathbb{Z}, n+1)\ \mathbb{Q}) = 0.$ is an isomorphism. A simple application of the Serre spectral sequence shows that this implies $\tilde{H}^*(K(\mathbb{Z}/k\,\mathbb{Z}, n); \mathbb{Q}) = 0$.

Step 2: $\pi \otimes \mathbb{Q} = 0 \Rightarrow \tilde{H}_*(K(\pi, n); \mathbb{Q}) = 0$.

If π is a sum of cyclic groups, then the result follows from Step 1. In general π is a direct limit of sums of cyclic groups. Since homology commutes with direct limits, Step 2 is true for all groups π.

Step 3: If $\pi_i(X) \otimes \mathbb{Q} = 0$ for all i and $\pi_1(X) = 0$, then $\tilde{H}_i(X; \mathbb{Q}) = 0$.

We build the Postnikov tower for X and prove the result inductively for the X_n. Suppose we know that $\tilde{H}_*(X_{n-1}; \mathbb{Q}) = 0$. Consider the Serre spectral sequence for $K(\pi_n(X), n) \to X_n \to X_{n-1}$ with $\mathbb{Q}$-coefficients. $E^2_{p,q} = 0$ for all $(p, q) \neq$ (0,0). Hence, $E^\infty_{p,q} = 0$ for all $(p, q) \neq$ (0,0). Thus, $\tilde{H}_i(X_n; \mathbb{Q}) = 0$ for all i. Since $H_i(X_n) = H_i(X)$ if $n > i$, this establishes Step 3.

Step 4: If $\pi_1(X) = 0$ and $\tilde{H}_i(X; \mathbb{Q}) = 0$ for all i, then $\pi_i(X) \otimes \mathbb{Q} = 0$ for all i.

Consider the Serre spectral sequence for

$$\begin{array}{ccc} K(\pi_n(X), n) & \longrightarrow & X_n \\ & & \downarrow \\ & & X_{n-1} \end{array}$$

for homology with $\mathbb{Q}$-coefficients. At the E^2-term, we have

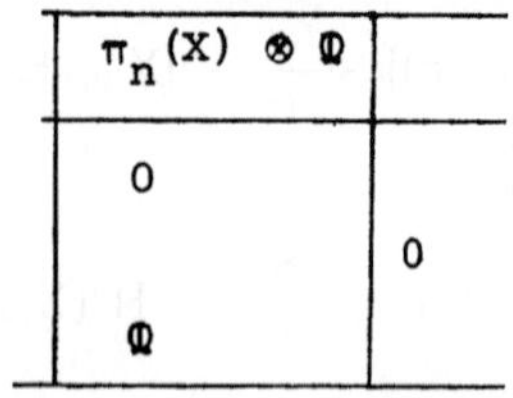

Thus, $\pi_n(X) \otimes \mathbb{Q} = E^{\infty}_{0,n} = H_n(X;\mathbb{Q})$. But $H_n(X_n) \cong H_n(X)$ and by assumption $H_n(X;\mathbb{Q}) = 0$. Thus, $\pi_n(X) \otimes \mathbb{Q} = 0$. This completes the lemma. ■

From this we prove (a) and (b) are equivalent. Namely, let $f\colon X \to Y$ and let F_f be the homotopy theoretic fiber. Then $f_*\colon \pi_*(X) \otimes \mathbb{Q} \to \pi_*(Y) \otimes \mathbb{Q}$ is an isomorphism $\Leftrightarrow$ $\pi_*(F_f) \otimes \mathbb{Q} = 0$ if and only if $H_*(F_f;\mathbb{Q}) = 0$ if and only if $H_*(X;\mathbb{Q}) \xrightarrow{f_*} H_*(Y;\mathbb{Q})$ is an isomorphism. (This last equivalence comes from the Serre spectral sequence.) This completes the proof of the theorem. ■

Definition. Given X and $f\colon X \to X_{(0)}$ with $X_{(0)}$ a $\mathbb{Q}$-space and f satisfying (1), (2), or (3) above (and hence all of them), we call $f\colon X \to X_{(0)}$ the *localization at* 0 of X.

The terminology comes from the fact that if we localize $\mathbb{Z}$ at 0 we get $\mathbb{Q}$.

We will not, in this course, consider any other localization, although it is possible to localize at any prime ideal. In fact, there is a Hasse–Minkowski principle which allows one to recover the whole space from its various localizations.

Theorem 8.9. *If* $\varphi\colon X \to X_{(0)}$ *and* $\varphi'\colon X \to X'_{(0)}$ *are localizations of* X*, then there is a homotopy equivalence* $h\colon X_{(0)} \to X'_{(0)}$ *such that*

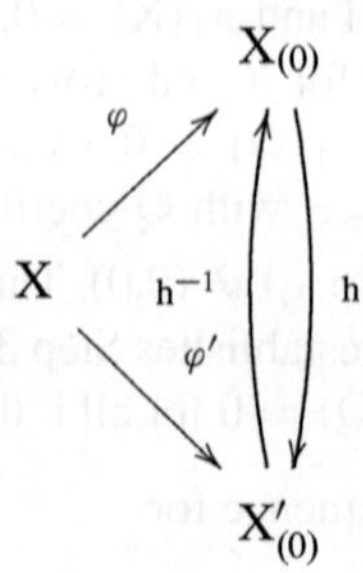

is a homotopy commutative diagram. Moreover, h *is unique up to homotopy.*

Proof. The proof follows immediately from the universal property of localization. ■

8.2 Construction of the Localization of a Space

The construction of the localization of a space goes by induction on the Postnikov tower of the space. We will assume that X is a CW complex and is simply connected. The idea of the proof is to tensor both the groups and the k-invariants with $\mathbb{Q}$.

Suppose inductively that we have a localization $X_{n-1} \xrightarrow{\varphi_{n-1}} (X_{n-1})_0$. Then

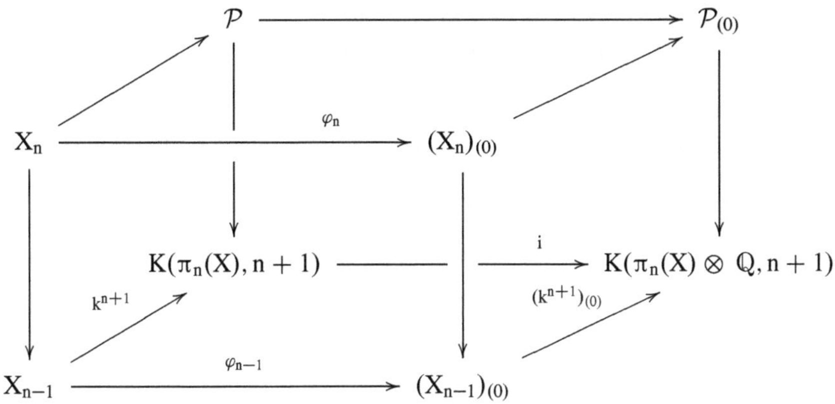

Since $K(\pi_n(X) \otimes \mathbb{Q}, n+1)$ is a $\mathbb{Q}$-space, $i \circ k^{n+1}$ factors uniquely through $(X_{n-1})_{(0)}$. Let $(X_n)_{(0)}$ be the fibration induced over $(X_{n-1})_{(0)}$ from $(k^{n+1})_{(0)}$. This defines $\varphi_{(n)}: X_{(n)} \to (X_n)_{(0)}$. By the commutativity of the diagram, we see that $(\varphi_{(n)})_*: \pi_n(X) \to \pi_n((X_n)_{(0)})$ is an isomorphism when tensored with $\mathbb{Q}$. This proves that $\varphi_{(n)}: X_{(n)} \to (X_n)_{(0)}$ is the localization of $X_{(n)}$ at 0. To complete the proof, we need the following lemma:

Lemma 8.10. *Let* Y *be a topological space. There is a CW complex* X *and a map* $\psi: X \to Y$ *which induces an isomorphism on all homotopy groups. If* $\psi': X' \to Y$ *is another such, then there is a homotopy equivalence* $h: X \to X'$ *so that* $\psi' \circ h$ *is homotopic to* ψ.

The proof of this lemma is left as an exercise for the reader.

We apply the lemma with $Y = \varprojlim (X_n)_{(0)}$. Let $X_{(0)}$ be a CW complex with $\psi: X_{(0)} \to Y$. The maps $X_n \to (X_n)_{(0)}$ define a map $\varprojlim X_n \to \varprojlim (X_n)_{(0)}$. Thus, we have

$$\begin{array}{ccccc} & & & & X_{(0)} \\ & & & & \downarrow \psi \\ X & \xrightarrow{\{\varphi_n\}} & \varprojlim X_n & \longrightarrow & \varprojlim (X^n)_{(0)} \end{array}$$

The obstructions to lifting the map $X \to \varprojlim X_{(0)}^{(n)}$ to a map $X \to X_{(0)}$ lie in $H^*(X; \pi_{*-1}(F))$ where F is the homotopy theoretic fiber of ψ. Since ψ induces an isomorphism on homotopy groups, $\pi_*(F) = 0$.

The resulting map $\varphi: X \to X_{(0)}$ is a localization at 0.

Calculations:

1. We have

$$H_*(K(\mathbb{Q}, 2n-1); \mathbb{Z}) = \begin{cases} \mathbb{Q} & \text{if } * = 2n-1 \\ 0 & \text{otherwise.} \end{cases}$$

Thus, the map $S^{2n-1} \to K(\mathbb{Q}, 2n-1)$ is a localization at 0, and consequently

$$\pi_*\left(S^{2n-1}\right) \otimes \mathbb{Q} \cong \begin{cases} \mathbb{Q} & \text{if } * = 2n-1 \\ 0 & \text{otherwise.} \end{cases}$$

Since $\pi_i(S^{2n-1})$ is always finitely generated

$$\pi_i(S^{2n-1}) \cong \begin{cases} \mathbb{Z} & \text{it } i = 2n-1 \\ \text{finite group} & \text{for } i \neq 2n-1. \end{cases}$$

2. $S^{2n} \to K(\mathbb{Q}, 2n)$ induces an isomorphism in rational cohomology through degree $(4n-1)$. The kernel of $H^{4n}(K(\mathbb{Q}, 2n); Q) \to H^{4n}(S^{2n}; \mathbb{Q})$ is generated by ι^2. Form the principle fibration $K(\mathbb{Q}, 4n-1) \to E \to K(\mathbb{Q}, 2n)$ with k-invariant ι^2. A calculation using the Serre spectral sequence shows that $H^*(E; \mathbb{Q}) \cong H^*(S^{2n}; \mathbb{Q})$. Furthermore, we can lift $S^{2n} \to K(\mathbb{Q}, 2n)$ to a map $S^{2n} \to E$. This proves that

$$\pi_i(S^{2n}) = \begin{cases} \mathbb{Z} & \text{for } i = 2n \\ \mathbb{Z} \oplus \text{finite group} & \text{for } i = 4n-1 \\ \text{finite group} & \text{otherwise.} \end{cases}$$

3. There is a map $\mathbb{C}P^n \to K(\mathbb{Q}, 2)$ which induces an isomorphism on rational cohomology through degree $2n+1$. Let $E \to K(\mathbb{Q}, 2)$ be a principal fibration with fiber $K(\mathbb{Q}, 2n+1)$ and k-invariant ι^{n+1}. As before one sees that $\mathbb{C}P^n \to E$ induces an isomorphism on rational cohomology. Thus, E is $(\mathbb{C}P^n)_{(0)}$.
4. We have seen that $H^*(BU(n); \mathbb{Q})$ is a polynomial algebra generated by $c_1, \ldots, c_n$. Define

$$(c_1, \ldots, c_n): BU(n) \longrightarrow \Pi_{k=1}^n K(\mathbb{Q}, 2k).$$

This map induces an isomorphism on rational cohomology. Thus, $BU(n)_{(0)} \cong \prod_{K=1}^{n} K(\mathbb{Q}, 2k)$. This means that all the k-invariants of BU(n) are trivial when tensored with $\mathbb{Q}$. This implies that they are all of finite order. This also shows the rational *Bott periodicity theorem* for the direct limit BU of the $BU(n)$, namely, $\Omega^2 BU_{(0)} \cong BU_{(0)}$.

These examples illustrate a recurring theme—homotopy theory over $\mathbb{Q}$ is much simpler than homotopy theory over $\mathbb{Z}$. We have a chance of getting complete answers over $\mathbb{Q}$, whereas over $\mathbb{Z}$, this is seldom possible; e.g., $\pi_*(S^n)$.

Chapter 9
deRham's Theorem for Simplicial Complexes

In the next four chapters, we will see how to calculate the rational homotopy type of a space from a certain differential graded algebra of forms.

9.1 Piecewise Linear Forms

We will work on a simplicial complex K. Recall that K is the union of simplices $\Delta^n \subset \mathbb{R}^{n+1}$, here Δ^n may be thought of as

$$\Delta^n = \left\{(t_0, \dots, t_n) \mid 0 \le t_i \le 1,\ \Sigma_{i=0}^n t_i = 1\right\}.$$

(The t_i are called *barycentric coordinates*.) The first few pictures are

0-simplex

1-simplex

2-simplex

3-simplex .

Note that $\partial\Delta^n$ is a union of $(n-1)$-simplices and is homeomorphic to S^{n-1}.

Consider the restriction to Δ^n of all forms in $\mathbb{R}^{n+1}$ of the form

$$\Sigma \varphi_{i_1 \dots i_j}\, dt_{i_j} \wedge \dots \wedge dt_{i_j}$$

where the $\varphi_{i_1 \dots i_j}$ are polynomials in the t_i with $\mathbb{Q}$-coefficients. There is the relation $\Sigma_{i=0}^n t_i = 1$, and the derived relation is $dt_0 + \dots + dt_n = 0$. We call this

P. Griffiths and J. Morgan, *Rational Homotopy Theory and Differential Forms*, Progress in Mathematics 16, DOI 10.1007/978-1-4614-8468-4_9,
© Springer Science+Business Media New York 2013

differential graded algebra, the algebra of *polynomial forms with* $\mathbb{Q}$*-coefficients* on Δ^n, and we denote it $A^*(\Delta^n)$. If $\Delta^k \subset \Delta^n$ is a face, then there is a restriction map $A^*(\Delta^n) \to A^*(\Delta^k)$.

Let K be a simplicial complex. Define the differential graded algebra of *piecewise polynomial forms with* $\mathbb{Q}$*-coefficients*, $A^*(K)$, to be all collections $\{(\omega_\sigma)_{\sigma \in K}$ that are *compatible* in the sense that if $\iota: \tau \to \sigma$ is the inclusion of a face, then $\iota^* \omega_\sigma = \omega_\tau$. Thus, $A^*(K)$ is the collection of forms, one on each simplex of K, which are compatible under restriction to faces. Clearly, wedge product and d, both defined by the corresponding operations on each simplex, give $A^*(K)$ the structure of a DGA defined over $\mathbb{Q}$).

Let $C^*(K; \mathbb{Q})$ denote the simplicial cochain complex of K. There is a map

$$\rho: A^*(K) \longrightarrow C^*(K;\ \mathbb{Q})$$

defined by $< \rho(\omega), \Delta^n >= \int_{\Delta_n} \omega$. This map is a map of cochain complexes by Stokes' theorem (which is valid in our setting).

Theorem 9.1 (p.l. deRham Theorem). ρ *induces an algebra isomorphism on cohomology.*

We will deduce the p.l. deRham theorem from the following proposition:

Proposition 9.2.

(i) *Let* $\varphi \in A^n(K)$ *satisfy* $d\varphi = 0$, $\rho(\varphi) = 0$ *(i.e.,* $\int_{\Delta_n} \varphi = 0$ *for all* Δ^n*). Then, there exists* $\psi \in A^{n-1}(K)$ *such that* $d\psi = \varphi$, $\rho(\psi) = 0$.

(ii) $A^*((K) \xrightarrow{p} C^*(K)$ *is onto.*

Proof of p.l. deRham Theorem Assuming Proposition 9.2 (Additive Statement): Let $B^*(K)$ be the kernel of ρ. By (ii) we have a short exact sequence of cochain complexes

$$0 \longrightarrow B^*(K) \longrightarrow A^*(K) \xrightarrow{\ \rho\ } C^*(K) \longrightarrow 0$$

leading to a long exact sequence of cohomology groups. The first part of the proposition says that

$$H^*\ (B^*(K)) = 0,$$

and thus, by the five lemma, it follows that

$$H^*_{dR}(K) \cong H^*(K;\ \mathbb{Q}).$$

The multiplicative statement will be considered later.

We turn now to the proof of Proposition 9.2.

9.2 Lemmas About Piecewise Linear Forms

Lemma 9.3 (Poincaré Lemma). *Let* $c(K)$ *be the simplicial complex that is the cone over a finite complex* K. *Suppose* φ^ℓ *is a closed form in* $A^*(c(K))$, $\ell > 0$. *Then,* $\varphi^\ell = d\psi^{\ell-1}$ *for some* $\psi^{\ell-1} \in A^*(c(K))$.

Proof. $c(K)$ is the join of the point c with K. Points in $c(K)$ are denoted by $sk + (1-s)\cdot c$ where $0 \le s \le 1$ and $k \in K$, with the proviso that $0\cdot k + 1\cdot c = c$ for all $k \in K$.

Define

$$\mu : c(K) \times I \to c(K)$$

by

$$\mu\,(s\cdot k + (1-s)\cdot c, t) = s(1-t)\cdot k + (1-s+st)\cdot c.$$

If $\lambda \in A^*(c(K))$, then $\mu^*(\lambda)$ is a form on $c(K)\times I$. Restricted to any $c(\sigma)\times I$, for σ a simplex of K, we have $\mu^*(\lambda)$ is a polynomial form with $\mathbb{Q}$ coefficients when expressed in terms of the barycentric coordinates, t_i, of σ and the coordinate s on I. These forms are compatible.

Claim. If φ^ℓ is a closed form of degree $\ell > 0$ on $c(K)$, then

$$d(-\int_{t=0}^{t=1} \mu^*(\varphi^\ell)) = \varphi^\ell.$$

Explanation: If we expand $\mu^*(\varphi^\ell)|c(\sigma)\times I$ as $\Sigma_{i=1}^N \alpha_i(\sigma)t^i + \beta_i(\sigma)t^i\,dt$ where $\alpha_i(\sigma)$ and $\beta_i(\sigma)$ are in $A^*(c(\sigma))$, then the $\alpha_i(\sigma)$ and $\beta_i(\sigma)$ patch together to give forms α_i and β_i in $A^*(c(K))$. If

$$\mu^*\varphi^\ell = \Sigma_{i=0}^N\,\alpha_i t^i + \beta_i t^i dt \text{with} \alpha_i, \beta_i \in A^*(c(K)),$$

then we define

$$\int_{t=0}^{t=1} \mu^*\varphi^\ell = \Sigma_{i=0}^N\,(-1)^{\deg \beta_i}\frac{\beta_i}{i+1}.$$

We now check that

$$d\left(-\int_{t=0}^{t=1}\mu^*\varphi\right) = \varphi.$$

Since $\mu|_{t=0}$ is the identity

$$\varphi = \mu^*\varphi|_{\substack{t=0\\ dt=0}} = \alpha_0.$$

Since $\mu|_{t=1}$ is the constant map to c,

$$\mu^*\varphi|_{\substack{t=1\\dt=0}} = \Sigma_{i=0}^N \alpha_i = 0.$$

(Here is the only place we use $\ell > 0$.)

Lastly, since φ is closed, $\rho^*(d\varphi) = d\mu^*\varphi = 0$; i.e.,

$$\begin{cases} d\alpha_i = 0 & \text{for } i \geq 0, \text{ and} \\ (-1)^{\deg \alpha_i} i\alpha_i + d\beta_{i-1} = 0 & \text{for } i \geq 1 \end{cases}$$

Hence,

$$d(-\int_{t=0}^{t=1} \rho^*\varphi) = \sum_{i=0}^{N} \left(\frac{(-1)^{\deg \beta_i + 1}}{i+1} \right) d\beta_i$$

$$= -\sum_{i=1}^{N} \alpha_i = \alpha_0 = \varphi.$$

■

Lemma 9.4 (Extension Lemma). *Let* φ^ℓ *be a form in* $A^*(\partial\Delta^n)$. *There is a form* $\psi^\ell \in A^*(\Delta^n)$ *such that* $\psi^\ell|\partial\Delta^n = \varphi$.

Proof. Let σ be an $(n-1)$-dimensional face of Δ^n, say $\sigma = \{(t_0, \ldots, t_n)|t_n = 0\}$. Let $\alpha \in A^*(\sigma)$. Let v be the vertex $\{t_n = 1\}$ and $U \subset \Delta^n$ be the complement of this vertex. There is a stereographic projection from the vertex $p: U \to \sigma$

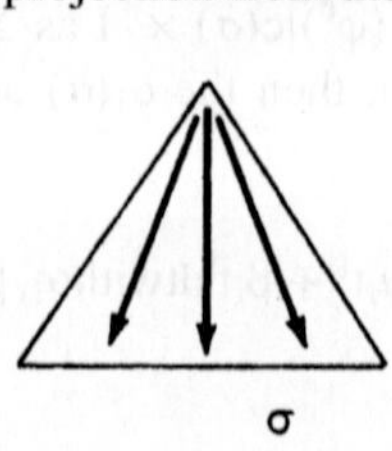

$$p(t_0, \ldots, t_n) = (\frac{t_0}{1-t_n}, \frac{t_1}{1-t_n}, \ldots, \frac{t_{n-1}}{1-t_n}).$$

The form $p^*(\alpha)$, considered as a form on U, is a polynomial form with $\mathbb{Q}$-coefficients in the variables $t_0, \ldots, t_{n-1}, 1/(1-t_n), dt_0, \ldots, dt_{n-1}, d(1/1-t_n)$. Since $d(1/1-t_n) = 1/(1-t_n)^2 dt_n$ the form $p^*(\alpha)$ is a polynomial form with $\mathbb{Q}$-coefficients in the variables. $t_0, \ldots, t_{n-1}, 1/1-t_n, dt_0, \ldots, dt_n$. Hence, for some $N \geq 0$, the form $(1-t_n)^N p^*(\alpha) = \tilde{\alpha}$ is a $\mathbb{Q}$-polynomial form on Δ^n. It is the required extension of α to all of Δ^n. Note that if τ is a face of σ and if $\alpha|\tau = 0$, then $\tilde{\alpha}$ restricted to the join of τ and v is 0.

Now suppose that $\varphi \in A^*(\partial\Delta^n)$. Let σ_0 be the face $\{t_0 = 0\}$, and let $\varphi_0 \in A^*(\sigma_0)$ be the restriction of φ to σ_0. Extend this to a form ψ_0 in Δ^n. The difference $\varphi - \psi_0|\partial\Delta^n$ vanishes on σ_0. Call this difference φ_1. Let σ_1 be the face $\{t_1 = 0\}$. Extend $\varphi_1|\sigma_1$ to a form ψ_1 on Δ^n. We know that $\psi_1|\sigma_0 = 0$. Thus, $\varphi-(\psi_0+\psi_1|\partial\Delta^n)$ vanishes on $\sigma_0 \cup \sigma_1$. Continue in this manner defining φ_i on and ψ_i on Δ^n In the end, we have $\varphi = \sum_{i=0}^n \psi_i|\partial\Delta^n$. ■

Lemma 9.5.

(a_n) *Let* φ^ℓ *be a closed form in* $A^*(\Delta^n)$ *which vanishes on* $\partial\Delta^n$. *If* $\ell = n$, *then assume also that* $\int_{\Delta_n} \varphi = 0$. *Then,* $\varphi^\ell = d\psi^{\ell-1}$ *for some* $\psi^{\ell-1}$ *which vanishes on* $\partial\Delta^n$.

(b_n) *Let* φ^ℓ *be a closed form in* $A^*(\partial\Delta^n)$, $\ell > 0$. *If* $\ell = n-1$, *then assume that* $\int_{\partial\Delta_n} \varphi = 0$. *Then,* $\varphi = d\psi$ *for some* $\psi \in A^*(\partial\Delta^n)$.

Proof. Clearly, (a_0) is true, and (b_0) and (b_1) are vacuous.
Proof of (a_1). Since k forms on a n-simplex vanish for $k > n$, to prove (a_1), we need only consider functions and 1-forms. (a_1) is clearly true for functions. For 1-forms, it is just the fundamental theorem of calculus. If $p(t)dt$ is a closed polynomial 1-form on $[0, 1]$, then there is a polynomial $q(t)$ so that $q'(t) = p(t)$ and $q(0) = 0$. Its value at 1 is $\int_0^1 p(t)\,dt$. Thus, if the integral of $p(t)dt$ over the interval is zero, then $q(t)$ vanishes on the boundary of the interval.
Proof that (a_{n-1}) *implies* (b_n). Let φ be a closed form in $A^*(\partial\Delta^n)$. Let $\sigma_n = \{t_n = 0\}$ be a codimension 1 face of Δ^n. By the Poincaré Lemma $\varphi|(\partial\Delta^n - \text{int}\,\sigma_n) = d\psi$ for some $\psi \in A^*(\partial\Delta^n - \text{int}\,\sigma_n)$. By the extension lemma, we can extend ψ to $\tilde\psi$ in $A^*(\partial\Delta^n)$. Then $\varphi - d\tilde\psi$ is a closed form on $\partial\Delta^n$ vanishing except on the face σ given by the equation $\{t_n = 0\}$. On this face, it is a relative form; i.e., it vanishes when restricted to $\partial\sigma^{n-1}$. If $\ell = n-1$, then

$$0 = \int_{\partial\Delta^n} \varphi^n = \int_{\partial\Delta^n} (\varphi^n - d\tilde\psi) = \int_{\sigma^{n-1}} (\varphi - d\psi).$$

Thus, by (a_{n-1}), we have $\varphi - d\tilde\psi = d\mu$ for some relative form μ on $\{t_n = 0\}$. Let $\tilde\mu$ be the extension by 0 of μ to the rest of $\partial\Delta^n$. Then, $\varphi = d(\tilde\psi + \tilde\mu)$.
Proof that (b_n) *implies* (a_n). Let φ^ℓ be a closed form on Δ^n with $\varphi^\ell|\partial\Delta^n = 0$. By the Poincaré lemma, $\varphi^\ell = d\psi^{\ell-1}$ for some form $\psi^{\ell-1}$ which may not vanish on $\partial\Delta^n$. Clearly $\psi^{\ell-1}|\partial\Delta^n$ is closed. If $\ell = 1$ and $n \geq 2$, then $\psi^{\ell-1}|\partial\Delta^n$ is a constant, and hence, by subtracting this constant, we can make $\psi^{\ell-1}|\partial\Delta^n = 0$. If $\ell > 1$, then by (b_n), $\psi^{\ell-1}|\partial\Delta^n = d\mu^{\ell-2}$. (If $\ell = n$, then

$$\int_{\partial\Delta^n} \psi^{\ell-1} = \int_{\Delta^n} d\psi = \int_{\Delta^n} \varphi = 0,$$

which allows us to apply (b_n).) Extend μ to a form $\tilde\mu$ on all of Δ^n. Then, $\varphi = d(\psi - d\tilde\mu)$ and $(\psi - d\tilde\mu)|\partial\Delta^n = 0$. ■

This completes the proof of the lemmas. Let us return to the two statements in Proposition 9.2.

(i) Suppose $\varphi \in A^n(K)$ satisfies $d\varphi = 0$ and $\int_{\sigma^n} \varphi = 0$ for all n-simplices $\sigma^n \in K$. Then, $\varphi = d\psi$ for some $\psi \in A^{n-1}(K)$ with the property that $\int_{\tau^{n-1}} \psi = 0$ for all $(n-1)$-simplices $\tau^{n-1} \in K$.

Proof. Given φ as above there is, according to Lemma 8.5, a form $\psi_\sigma \in A^{n-1}(\sigma^n)$ such that $\varphi|\sigma = d\psi_\sigma$ and such that $\psi_\sigma|\partial\sigma^n = 0$. The collection $\{\psi_{\sigma^n}\}$ defines an element in $A^{n-1}(K^{(n)})$ which vanishes on $K^{(n-1)}$. Using the extension lemma repeatedly, we can extend this form to ψ_n on all of K. Clearly $\varphi - d\psi_n$ vanishes on $K^{(n)}$. The difference $(\varphi - d\psi_n)|\sigma^{n+1}$ is closed, and for every $(n+1)$-simplex σ^{n+1}, this form vanishes on $\partial\sigma^{n+1}$. Hence, there is a form $\mu_\sigma \in A^{n-1}(\sigma^{n+1})$ such that $d\mu_\sigma = (\varphi - d\psi_n)|\sigma$ and $\mu_\sigma|\partial\sigma^{n+1} = 0$. As before, the $\{\mu_\sigma\}$ define a form on $K^{(n+1)}$ vanishing on $K^{(n)}$. This can be extended to a form ψ_{n+1} on K which vanishes on $K^{(n)}$. The difference $\varphi - d(\psi_n + \psi_{n+1})$ vanishes on $K^{(n+1)}$. Continuing in this manner, we define $\psi_n, \psi_{n+1}, \ldots, \psi_i \in A^{n-1}(K)$, such that $\psi_i|K^{(i-1)} = 0$ and $\varphi - d(\psi_n + \ldots + \psi_{n+k})$ vanishes on $K^{(n+k)}$. Because ψ_i vanishes on $K^{(i-1)}$, the infinite sum $\Sigma_{k=0}^{\infty} \psi_{n+k}$ is an element of $A^*(K)$. It is the required element ψ. ■

(ii) $A^*(K) \xrightarrow{\rho} C^*(K)$ is onto.

Proof. Given a simplex $\sigma^n \in K$, there is a form in σ^n, $(1/\text{vol}\,\sigma^n)\,(dt_1 \wedge \ldots \wedge dt_n)$, which has integral 1. This form can be extended by 0 to the rest of the n-skeleton of K. Using the extension lemma, we then extend it to all of K.

The result is an n-form φ such that

$$\int_{\sigma^n} \varphi = 1 \text{ and } \int_{\tau^n} \varphi = 0 \text{ for } \tau^n \neq \sigma^n.$$

Since we can do this for any simplex, the map ρ is onto. ■

9.3 Naturality Under Subdivision

Let K be a linear cell complex. This means that $K = UD_i$ where D_i is a convex linear cell. On the intersection of two cells, the linear structures agree. *A subdivision* of K is a new linear cell structure on K so that each cell in the new decomposition lies affine linearly in a cell of the old. Such subdivisions arise by choosing points which are to be the new vertices. If we choose exactly one point from each cell, the resulting subdivision is called *barycentric subdivision*. Barycentric subdivision always yields a simplicial complex.

Lemma 9.6. *Let* K *be a linear cell complex, and let* K′ *be a subdivision of* K *which is a simplicial complex. Then the integration map*

$$\rho\colon A^*(K') \to C^*(K; \mathbb{Q})$$

induces an isomorphism on cohomology.

Proof. The definition of ρ is to integrate forms in $A^*(K')$ on cells of K. Now apply the p.l. deRham theorem for each cell of K and argue by induction over the cells. ■

If K is a simplicial complex and K′ is a subdivision of K in which each new vertex has rational barycentric coordinates in K, then restriction induces a map of differential algebras defined over $\mathbb{Q}$,

$$A^*(K) \to A^*(K').$$

By the p.l. deRham theorem, this map is an isomorphism on cohomology.

If $f\colon |K| \to |L|$ is a continuous map between the geometric realizations of simplicial complexes, then it is possible to subdivide K sufficiently to simplicial complex K′, with the property that there is a simplicial map $\varphi\colon K' \to L$ homotopic to f. This approximation induces $|\varphi|\colon |K| \to |L|$ which is homotopic to f. Since K' could be any sufficiently fine subdivision of K, we can take it to be a rational subdivision. Hence, we will have $\varphi^*\colon A^*(L) \to A^*(K')$ resulting from a continuous map $f\colon |K| \to |L|$. Of course φ^* will depend significantly on many choices (i.e., which subdivision and which approximation we take), but (as we shall see) in the homotopy category of differential algebras φ^* is well defined.

9.4 Multiplicativity of the deRham Isomorphism

Let K be a simplicial complex, $A^*(K)$ the piecewise linear forms (p.l. forms) on K, and $H^*_{dR}(K)$ its cohomology. We have shown that

$$\rho^*\colon H^*_{dR}(K) \longrightarrow H^*(|K|; \mathbb{Q})$$

is an additive isomorphism. Both $H^*_{dR}(K)$ and $H^*(|K|; \mathbb{Q})$ are naturally graded rings: H^*_{dR} from wedge product of forms and $H^*(|K|; \mathbb{Q})$ from the Alexander–Whitney formula for cup product of cochains. We want to prove that ρ^* is an isomorphism of the graded cohomology algebras. The easiest path to this goal is to give a different description of the cup product in singular cohomology. Let K be a simplicial complex with underlying space |K|. The space $|K| \times |K|$ carries naturally the structure of a linear cell complex. The cells are all products $\sigma \times \tau$ with σ and τ simplices of K. This cell structure is not a simplicial complex structure. (If it were, then there would be a graded-commutative, associative cup product on simplicial cochains.) Let α and β be simplicial cochains on K. We define $\alpha \otimes \beta$ as a cellular cochain on $|K| \times |K|$. The value is given by $< \alpha \otimes \beta, \sigma \times \tau > = < \alpha, \sigma > \cdot < \beta, \tau >$. One sees easily that $\delta(\alpha \otimes \beta) = \delta\alpha \otimes \beta + (-1)^{\deg \alpha} \alpha \otimes \delta\beta$. Consequently, if α and β are cocycles, then so is $\alpha \otimes \beta$. The class $[\alpha \otimes \beta] \in H^*(|K| \times |K|))$, and when restricted to the diagonal $\Delta: |K| \to K| \times |K|$, gives $[\alpha] \cup [\beta]$ in $H^*(|K|)$.

The Alexander–Whitney formula for multiplication arises from choosing a homotopy from $\Delta: |K| \to |K| \times |K|$ to a linear mapping (called a chain approximation to the diagonal).

Let φ_0 and φ_1 be elements of $A^*(K)$. Let $(|K| \times |K|)'$ be a rational subdivision of $|K| \times |K|$ which is a simplicial complex. Define $\varphi_0 \otimes \varphi_1 \in A^*((|K| \times |K|)')$ as follows. Each simplex $\tau \subset (|K| \times |K|)'$ lies in a product $\sigma_0 \times \sigma_1$ in such a way that its vertices are rational. The form $(\varphi_0|\sigma_0) \otimes (\varphi_1|\sigma_1)$ is a polynomial form with rational coefficients in the product linear structure. Thus, $(\varphi_0|\sigma_0) \otimes (\varphi_1|\sigma_1)$ restricts to give a form $\varphi_0 \otimes \varphi_1(\tau) \in A^*(\tau)$. Clearly, these forms fit together to define $\varphi_0 \otimes \varphi_1 \in A^*((|K \times |K|)')$.

Under the map $\rho: A^*((|K| \times |K|)') \to C^*(|K| \times |K|; \mathbb{Q})$, the form $\varphi_0 \otimes \varphi_1$ goes to the cochain which evaluates on $\sigma_0 \times \sigma_1$ to give $(\int_{\sigma_0} \varphi_0) \cdot (\int_{\sigma_1} \varphi_1)$. Thus, if φ_0 and φ_1 are closed forms, then $\rho(\varphi_0 \otimes \varphi_1)$ is the cocycle $\rho(\varphi_0) \otimes \rho(\varphi_1)$. Restricting to the diagonal, we see that the singular cohomology class of $\varphi_0 \wedge \varphi_1$ is the cup product of the classes of φ_0 and φ_1. This proves that $\rho^*: H^*(A^*(K)) \to H^*(C^*(K); \mathbb{Q})$ is an algebra isomorphism.

9.5 Connection with the C^∞ deRham Theorem

If M is a C^∞ manifold, then associated to it is the differential algebra of C^∞ forms. It is an algebra over $\mathbb{R}$. The original theorem of deRham says that the cohomology of this differential algebra is naturally isomorphic (as a ring) to the singular cohomology with real coefficients. The connection between forms on singular cochains is once again achieved by integration. There are many proofs by now of deRham's theorem. For example, one can use currents to give a proof (essentially deRham's original proof); one can prove that the resulting homology groups satisfy the Eilenberg–Steen road axioms and hence must be singular homology. More in the spirit of the present discussion one can prove the Poincaré lemma and then establish the isomorphism by induction on a handlebody (instead of a triangulation). Similarly one could use a cover by convex subsets of $\mathbb{R}^n$, but in this setup, the induction is more complicated and goes under the appellation of "sheaf theory."

Whatever method is chosen, the result is the following, cf. [27]:

Theorem 9.7. *Let $A^*_{C^\infty}(M)$ denote the DGA of C^∞ forms on M. Then, a map of cochain complexes induced by integration*

$$\rho: A^*_{C^\infty}(M) \to C^*(M, \mathbb{R})$$

induces an isomorphism of cohomology rings.

Any C^∞ manifold has a C^∞ triangulation. This means that there is a simplicial complex K_M and a homeomorphism $\varphi\colon K_M \to M$ which is C^∞ on each simplex. Consider the DGA of $\mathbb{Q}$-polynomial forms $A^*(K_M)$. This gives a second algebra of forms on M. We wish to relate these. Let

$$A^*(M) = \left\{ \begin{array}{l} \text{piecewise polynomial forms on M with } \mathbb{Q}\text{-coefficients} \\ \text{for the given triangulation} \end{array} \right\}$$

$$A^*_{C^\infty}(M) = \{C^\infty - \text{forms}\}$$

$$\tilde{A}^*(M) = \left\{ \begin{array}{l} \text{collections } \{\varphi\} \text{ of forms on the} \\ \text{simplices } \{\Delta^n\} \text{ such that} \\ \text{(i) } \varphi \text{ is } C^\infty \text{ in } \bar{\Delta}^n \text{ for each n-simplex} \\ \text{(ii) } \varphi|\Delta^n \cap \Delta'^n = \varphi'|\Delta^n \cap \Delta'^n. \end{array} \right\}$$

Thus, $\tilde{A}^*(M)$ is constructed in the same way as $A^*(M)$ only now using C^∞ forms. This definition makes sense, and $\tilde{A}^*(M)$ is graded algebra having a "d." Also, Stokes' theorem for simplicial chains holds as before. We call forms in $\tilde{A}^*(M)$ *piecewise* C^∞*-forms*. We have the following inclusions:

$$\begin{array}{ccc} A^*(M) \otimes_{\mathbb{Q}} \mathbb{R} & & A^*_{C^\infty}(M) \\ & \searrow \quad \swarrow & \\ & \tilde{A}^*(M) & \end{array}$$

If we denote the cohomology of $\tilde{A}^*(M)$ by $H^*_{p.C^\infty}(M)$, then the integration map induces a map $H^*_{p.C^\infty}(M) \to H^*(M;\ \mathbb{R})$. Thus, we have the following commutative diagram of isomorphisms:

$$\begin{array}{ccccc} & & H^*_{p.C^\infty}(M) & & \\ & \nearrow & & \nwarrow & \\ H^*(A^*(M)) \otimes_{\mathbb{Q}} \mathbb{R} & & \downarrow & & H^*(A^*_{C^\infty}(M)) \\ & \searrow & & \swarrow & \\ & & H^*(M; \mathbb{R}) & & \end{array}$$

We shall show:

Proposition 9.8. *Integration induces an isomorphism* $H^*_{p.C^\infty}(M) \to H^*(M;\ \mathbb{R})$.

Corollary 9.9. *The inclusions* $A^*(M) \otimes_{\mathbb{Q}} \mathbb{R} \to \tilde{A}^*(M)$ *and* $A^*_{C\infty}(M) \to \tilde{A}^*(M)$ *both induce isomorphisms on cohomology.*

Proof (of Proposition 9.8). The proof is essentially the same as in the p.l. case. The Poincaré lemma holds and is proved the same way. The extension lemma is again proved using stereographic projection, but instead of multiplying by a power of $(1 - t_n)$, we simply multiply by a C^∞ function of t_n which is 1 for $t_n = 0$ and is identically 0 near $t_n = 0$. Once we have these lemmas, the argument given in the $\mathbb{Q}$-polynomial case is valid, mutatis mutandis, in the piecewise C^∞ case. ■

Corollary 9.9 follows immediately from the commutative diagram and Proposition 9.8.

9.6 Generalizations of the Construction

Suppose that X is a space made out of manifold "pieces." This includes, for example, a simplicial complex. Then it is possible to define an algebra of piecewise C^∞ forms on X which will calculate the cohomology. We don't prove this result in general, but rather confine ourselves to one important example.

Let $D_1, \dots, D_k$ be smooth submanifolds in an ambient manifold Y. Suppose that all intersections $D_{i_1} \cap \dots \cap D_{i_t}$, $i_1 < i_2 < \dots < i_t$ are transverse. We define $A^*_{p.C\infty}(\cup_{i=1}^k D_i)$ to be compatible collections of forms

$$\{\omega_i \in A^*_{C\infty}(D_i),\ i = 1, \dots, t,\ \text{satisfying } \omega_i|D_i \cap D_j = \omega_j|D_i \cap D_j \text{ for all } i, j\}.$$

We claim that the cohomology of $A^*_{p.C\infty}\left(\cup_{i=1}^k D_i\right)$ is isomorphic to the singular cohomology of $\cup_{i=1}^k D_i$ with real coefficients. Triangulate the union so that all intersections are subcomplexes and each simplex in the triangulation is C^1. Let K be the resulting simplicial complex. There is a map

$$A^*_{p.C\infty}(\cup_{i=1}^k D_i) \longrightarrow C^*(K; \mathbb{R})$$

given by integrating the form over the simplices of K. We claim that this map induces an isomorphism on cohomology. This is proved by induction on k. If $k = 1$, then this is exactly the deRham theorem. Suppose we have established the result for all $\ell < k$. We write $\cup_{i=1}^k D_i$ as $(\cup_{i=1}^{k-1} D_i) \cup D_k$ where the intersection is $\cup_{i=1}^{k-1}(D_i \cap D_k)$. Let $D' = \cup_{i=1}^{k-1} D_i$ and $D = \cup_{i=1}^k D_i$. We have a commutative diagram of short exact sequences

$$\begin{array}{ccccccccc}
0 & \longrightarrow & A^*_{p.C^\infty}(D) & \longrightarrow & A^*_{p.C^\infty}(D' \coprod D_k) & \longrightarrow & A^*_{p.C^\infty}(D' \cap D_k) & \longrightarrow & 0 \\
 & & \downarrow & & \downarrow & & \downarrow & & \\
0 & \longrightarrow & A^*_{C^\infty}(D) & \longrightarrow & A^*_{C^\infty}(D' \coprod D_k) & \longrightarrow & A^*_{C^\infty}(D' \cap D_k) & \longrightarrow & 0
\end{array}$$

The result follows immediately by induction and the five lemma.

Chapter 10
Differential Graded Algebras

10.1 Introduction

In this chapter, we shall study differential algebras in their own right. What we are doing, actually, is studying the homotopy theory of differential algebras. In fact, we shall construct an object (the minimal model) which should be considered the Postnikov tower of a differential algebra.

Definition. A *differential graded algebra* (or DGA for short), $\mathcal{A}^*$, is a graded vector space over $\mathbb{Q}$, $\mathbb{R}$, or $\mathbb{C}$,

$$\mathcal{A}^* = \oplus_{p \geq 0} \mathcal{A}^p,$$

having

(i) A differentiation $d: \mathcal{A}^* \to \mathcal{A}^{*+1}$ with $d^2 = 0$.
(ii) A multiplication $\mathcal{A}^p \otimes \mathcal{A}^q \to \mathcal{A}^{p+q}$ satisfying

$$\alpha\beta = (-1)^{pq}\beta\alpha.$$

(iii) $d(\alpha\beta) = d\alpha\beta + (-1)^p \alpha d\beta$.

Examples. (i) The C^∞ deRham complex $A^*_{dR}(M)$ of a smooth manifold and the p.l. deRham complex $A^*_{p.l.}(K)$ of a simplicial complex are DGAs over $\mathbb{R}, \mathbb{Q}$, respectively.

(ii) The cohomology $H^*(X, \mathbb{Q})$ of a space is a DGA (with $d = 0$), but the singular cochain complex $C^*(X, \mathbb{Q})$ is not (the signed commutativity fails).

(iii) The commutative cochain problem. Perhaps the main genesis of the theory we are considering is the problem of commutative cochains. This was solved in an abstract manner by Quillen, and in an attempt to better understand this, Sullivan was led to the p.l. forms and the connection between differential forms and homotopy type. In retrospect one can already see much of the

P. Griffiths and J. Morgan, *Rational Homotopy Theory and Differential Forms*,
Progress in Mathematics 16, DOI 10.1007/978-1-4614-8468-4_10,

© Springer Science+Business Media New York 2013

theory in Whitney's book [27]; however, one fundamental point was missing in that Whitney only constructs commutative cochains over $\mathbb{R}$, and as already mentioned there is no way to build Postnikov towers over $\mathbb{R}$ and thus tie in the commutative cochains with homotopy type.

Let X be a simplicial complex. The usual definition of the cup product $\alpha_\alpha \cup \beta_q$ between a p-cochain α_q and a q-cochain β_q is

$$< \alpha_p \cup \beta_q,\ \Delta^{p+q} >$$
$$=< \alpha_p,\ \text{front p-face of } \Delta^{p+q} > \cdot < \beta_q,\ \text{back q} - \text{face of } \Delta^{p+q} > .$$

This formula leads to the following properties:

(i) $\delta(\alpha_p \cup \beta_q) = \delta\alpha_p \cup \beta_q + (-1)^p \alpha_p \cup \delta\beta_q$.
(ii) $\alpha_p \cup (\beta_q \cup \gamma_r) = (\alpha_p \cup \beta_q) \cup \gamma_r$.

Moreover, a somewhat grizzly computation shows that on the cohomology level, we have graded commutativity

(iii) $[\alpha_p] \cup [\beta_q] = (-1)^{pq}[\beta_q] \cup [\alpha_p]$.

However, it is obviously false that

$$\alpha_p \cup \beta_q = (-1)^{pq}\beta_q \cup \alpha_p$$

on the cochain level. Now one may attempt to modify the formula so as to have all three properties, but all such attempts are doomed to failure since, as realized by Steenrod 35 years ago, the failure to find commutative cochains over $\mathbb{Z}$ is reflected in the existence of cohomology operations, such as the Steenrod squares. These objections do not apply over $\mathbb{Q}$ (we have essentially proved this by calculating $H^*(K(\mathbb{Z}, n), \mathbb{Q})$); thus, it is reasonable to look for commutative cochains/$\mathbb{Q}$. Before going on, it is time to precisely define what is meant by commutative cochains.

Definition. Commutative cochains assign functorially to each simplicial complex X a DGA defined over $\mathbb{Q}$, $C^*(X)$, satisfying (i)–(iii) above and such that:

(iv) The cohomology of $C^*(X)$ is $H^*(X; \mathbb{Q})$.
(v) Given a subcomplex $Y \subset X$, we have

$$C^*(X) \to C^*(Y) \to 0.$$

Example. The p.l. forms $A^*_{\text{p.l.}}(X)$ give an explicit solution to the commutative cochain problem. The cohomology $H^*(X, \mathbb{Q})$ does not give a solution because (v) is violated.

The argument given in this chapter shows that

Theorem. *Let* $C^*(X)$ *be any solution to the commutative cochain problem over* $\mathbb{Q}$. *Then the minimal model* $\mathfrak{M}$ *for* $C^*(X)$ *gives the* $\mathbb{Q}$*-homotopy type of* X.

So now we have some full circle. The problem of commutative cochains is equivalent to finding not only the cohomology but also the $\mathbb{Q}$-homotopy type of a space from a cochain complex. The p.l. forms explicitly solve this problem, and moreover, a simple comparison theorem shows that the C^{∞} forms give the $\mathbb{R}$-homotopy type of a smooth manifold.

It is interesting to note that in [27], Whitney essentially showed that any solution to the commutative cochain problem over $\mathbb{R}$ satisfying a mild continuity condition is given by integration of suitable differential forms (the flat forms) over chains. Now, almost 25 years later, we have finally understood what he was driving at.

Given a DGA, $\mathcal{A}^*$, we denote by $H^*(\mathcal{A}^*)$ the cohomology algebra. It is again a DGA with $d = 0$.

We assume throughout that $H^0(\mathcal{A}^*)$ is the ground field and that $H^1(\mathcal{A}^*) = 0$. Thus, $\mathcal{A}^*$ is, so to speak, *simply connected.*

Definition. A DGA $\mathcal{A}^*$ is said to be *minimal* if:

(i) $\mathcal{A}^*$ is free as a graded-commutative algebra on generators of degrees ≥ 2.
(ii) $d(\mathcal{A}^*) \subset \mathcal{A}^+ \wedge \mathcal{A}^+$ where $\mathcal{A}^* = \oplus_{k>0}\mathcal{A}^k$.

Condition (i) means that $\mathcal{A}^*$ is a tensor product of polynomial algebras on generators of even degrees and exterior algebras on generators of odd degrees. Condition (ii) says that d is *decomposable*. There is a notion of minimal DGAs which have generators in degree 1 (see Chap. 13).

Given a DGA, $\mathcal{A}^*$, we wish to construct a *minimal model*, $\mathfrak{M}(\mathcal{A}^*)$, for $\mathcal{A}^*$. By definition this means that $\mathfrak{M}(\mathcal{A}^*)$ is a minimal DGA and there is a map $\rho: \mathfrak{M}(\mathcal{A}^*) \to \mathcal{A}^*$ of DGAs inducing an isomorphism on cohomology. One of the main results of this chapter is that every simply connected DGA has a minimal model.

Remark. The construction of $\mathfrak{M}(\mathcal{A}^*)$ is motivated by the construction of the Postnikov tower of a space. In fact, the parallel is quite precise, as we shall see in Chap. 9.

10.2 Hirsch Extensions

Actually the fundamental property of a minimal algebra is that it is an increasing sequence of subalgebras which are nicely related, one to the next. To illuminate this, we study these extensions separately. First

Definition. Let $\mathcal{A}$ be a DGA. A *Hirsch extension* of $\mathcal{A}$ is an inclusion

$$\mathcal{A} \to \mathcal{A} \otimes_d \Lambda(V^k).$$

The notation means that (i) V is a (finite dimensional) vector space homogeneous of degree k; (ii) $\Lambda(V^k)$ is the free graded-commutative algebra with unit generated by V (the polynomial algebra on V if k is even and the exterior algebra on V if k is odd); and (iii) d: $V \to \mathcal{A}^{k+1}$. The differential on the full algebra is determined by $d|A$ and $d|V$.

Two Hirsch extensions $\mathcal{A} \to \mathcal{A} \otimes_d \Lambda(V^k)$ and $\mathcal{A} \to \mathcal{A} \otimes_d \Lambda((V')^k)$ are equivalent if there is a commutative diagram

$$\begin{array}{ccc} & & \mathcal{A} \otimes_d \Lambda(V^k) \\ & \nearrow & \\ \mathcal{A} & & \downarrow \varphi \\ & \searrow & \\ & & \mathcal{A} \otimes_{d'} \Lambda((V')^k) \end{array}$$

with φ an isomorphism which is the identity on $\mathcal{A}$.

Lemma 10.1. *$\mathcal{A} \to \mathcal{A} \otimes_d \Lambda(V)$ and $\mathcal{A} \to \mathcal{A} \otimes_{d]} \Lambda(V')$ are equivalent if and only if there is an isomorphism $\psi: V \to V'$ so that the following diagram commutes:*

$$\begin{array}{ccc} V & \xrightarrow{\ d\ } & H^{k+1}(\mathcal{A}) \\ \psi \downarrow & & \downarrow = \\ V' & \xrightarrow{\ d'\ } & H^{k+1}(\mathcal{A}) \end{array}$$

Proof. If $\varphi: \mathcal{A} \otimes_d \Lambda(V) \xrightarrow{\sim} \mathcal{A} \otimes_{d'} \Lambda(V')$ is an isomorphism extending the identity on A, then $\varphi(v) = a_v + \psi(v)$. For φ to be an isomorphism, ψ must be an isomorphism. Since $\varphi(dv) = d'(\varphi(v)) = d'a_v + d'\psi(v)$,

$$[dv] = [d'\psi(v)] \in H^{k+1}(\mathcal{A}).$$

Conversely, if we have $\psi: V \xrightarrow{\sim} V'$ so that $[dv] = [d'\psi(v)] \in H^{k+1}(\mathcal{A})$, then $dv - d'\psi(v) = da_v$ for some $a_v \in \mathcal{A}$. We can choose a_v linearly in v. Define $\varphi(v) = a_v + \psi(v)$. This defines a map $\varphi: \mathcal{A} \otimes \Lambda(V) \to \mathcal{A} \otimes \Lambda(V')$, which is easily seen to be an isomorphism and to commute with the differentials. ■

To classify Hirsch extensions of $\mathcal{A}$ with a fixed vector space of new generators, V, we say that two are equivalent if the isomorphism

$$\varphi: \mathcal{A} \otimes_d \Lambda(V) \longrightarrow \mathcal{A} \otimes_{d'} \Lambda(V)$$

is the identity on A and sends v to $a_v + v$. Equivalence classes are then in natural one-to-one correspondence with maps

$$d: V \longrightarrow H^{k+1}(\mathcal{A});$$

or what is the same thing, the class of

$$[d] \in H^{k+1}(\mathcal{A}; V^*).$$

Proposition 10.2. *Let $\mathfrak{M}$ be a DGA and let $\mathfrak{M}(n) \subset \mathfrak{M}$ be the subalgebra generated in degrees $\leq n$. Then, $\mathfrak{M}$ is a minimal DGA if and only if*

$$\begin{aligned} \mathfrak{M}(1) &= \textit{ the ground field} \\ \mathfrak{M}(0) &= \mathfrak{M}(1) \subset \mathfrak{M}(2) \subset \mathfrak{M}(3) \subset \dots \\ \cup_n \mathfrak{M}(n) &= \mathfrak{M}, \end{aligned}$$

and for each n *the inclusion $\mathfrak{M}(n) \subset \mathfrak{M}(n+1)$ is a Hirsch extension with the new generators being in degree* $n+1$.

Proof. Suppose that $\mathfrak{M}$ is minimal. Then clearly, $\mathfrak{M}(1)$ is the ground field and $\mathfrak{M} = \cup_n \mathfrak{M}(n)$. Since each $\mathfrak{M}(i)$ is free as a graded-commutative algebra, it is clear that as vector spaces

$$\mathfrak{M}(n+1) \cong \mathfrak{M}(n) \otimes_d \Lambda\,(V^{n+1}).$$

Since $d(v)$ is decomposable for $v \in V^{n+1}$ and $\mathfrak{M}$ has no elements of degree 1, it must be the case that $d(v)$ is a sum of products of elements of degree $\leq n$; i.e., $d(v) \in \mathfrak{M}(n)$. This proves that $\mathfrak{M}(n) \subset \mathfrak{M}(n+1)$ is a Hirsch extension.

Conversely, if $\mathfrak{M} = \cup_n \mathfrak{M}(n)$ where $\mathfrak{M}(n) \subset \mathfrak{M}(n+1)$ is a Hirsch extension of degree $n+1$ and $\mathfrak{M}(1)$ is the ground field, then by induction, one see that $\mathfrak{M}(n)$ is a minimal DGA. It follows that $\mathfrak{M}$ is also a minimal DGA. ■

10.3 Relative Cohomology

Before beginning the actual construction of $\mathfrak{M}(\mathcal{A}^*)$, we need a few basic facts about relative cohomology for a map between two cochain complexes. Let $f: C^* \to D^*$ denote a degree-preserving map between two cochain complexes. Define

$$M_f^n = C^n \oplus D^{n-1}$$

and let $\delta: M_f^n \to M_f^{n+1}$ be given by

$$\begin{pmatrix} \delta_C & 0 \\ f & -\delta_D \end{pmatrix}.$$

One checks easily that $\delta^2 = 0$. We define $H^*(C, D)$ to be $H^*(M_{f^*})$. The maps $(-i_2)\colon D^{*-1} \longrightarrow M_f^*$ and $\pi_1\colon M_f^* \longrightarrow C^*$ commute with the coboundaries. The resulting maps on cohomology give a long exact sequence

$$\longrightarrow H^n(C) \xrightarrow{f^*} H^n(D) \xrightarrow{\delta} H^{n+1}(C, D) \xrightarrow{i^*} H^{n+1}(C) \xrightarrow{f^*} \ldots.$$

Clearly, this long exact sequence is functorial for commutative diagrams

$$\begin{array}{ccc} C_1 & \xrightarrow{f_1} & D_1 \\ \downarrow & & \downarrow \\ C_2 & \xrightarrow{f_2} & D_2. \end{array}$$

10.4 Construction of the Minimal Model

Theorem 10.3. *Every simply connected DGA has a minimal model.*

Proof. Given a DGA $\mathcal{A}$, we shall construct an increasing sequence of Hirsch extensions

$$\text{ground field} = \mathfrak{M}(0) = \mathfrak{M}(1) \subset \mathfrak{M}(2) \subset \ldots$$

together with maps

$$\rho_n\colon \mathfrak{M}(n) \longrightarrow \mathcal{A}$$

so that

$$\rho_n|\mathfrak{M}(k) = \rho_k \text{for} \quad k \leq n$$

and the map $\rho_n\colon \mathfrak{M}(n) \to \mathcal{A}$ is an n-*minimal model* in the sense that:

(i) $\mathfrak{M}(n)$ is minimal and generated by elements in degrees $\leq$ n.
(ii) ρ_n^* is an isomorphism on cohomology in degrees $\leq$ n.
(iii) ρ_n^* is an injection on cohomology in degree $n + 1$.

To begin with, $\mathfrak{M}(1)$ is the ground field and ρ_1 is the map sending 1 to 1. Obviously, this map is an isomorphism on cohomology in degrees ≤ 1 and an injection on cohomology in degree 2. Suppose for some $n \geq 1$ that we have inductively constructed $\mathfrak{M}(n)$ and $\rho_n\colon \mathfrak{M}(n) \to \mathcal{A}$ as required. The relative cohomology $H^i(\mathfrak{M}(n), \mathcal{A}))$ vanishes for $i \leq n+1$. (This follows from the long exact sequence of the pair and the conditions on ρ_n.) Let $V = H^{n+2}(\mathfrak{M}(n), \mathcal{A})$. We will form $\mathfrak{M}(n+1)$ by taking the algebra

$$\mathfrak{M}(n) \otimes \Lambda(V^{n+1}).$$

To extend the differential on $\mathfrak{M}(n)$ to one defined on all of $\mathfrak{M}(n+1)$, by the Leibnitz rule, we need only define it on V subject to the condition that $dv \in \mathcal{A}^{n+2}$ is closed for all $v \in V$. Having done this, to define a map $\rho_{n+1}: \mathfrak{M}(n+1) \to \mathcal{A}$ extending ρ_n, it is necessary only to define $\rho_{n+1}|V$ subject to the condition

$$\rho_n(dv) = d\rho_{n+1}(v) \text{ for all } v \in V.$$

Both d and ρ_{n+1} are defined by a linear splitting s for the projection map from cocycles to cohomology:

$$Z^{n+2}(\mathfrak{M}, \mathcal{A}) \underset{s}{\overset{\longrightarrow}{\longleftarrow}} H^{n+2}(\mathfrak{M}, \mathcal{A})$$

This is equivalent to choosing, linearly in v, cocycle representatives $(m_v, a_v) \in \mathfrak{M}(n)^{n+2} \oplus \mathcal{A}^{n+1}$ for the relative cohomology class v. For (m_v, a_v) to be a cocycle means $dm_v = 0$ and $\rho_n(m_v) = da_v$. Having chosen the splitting from relative cohomology back to the relative cocycles, we let

$$d(v) = m_v$$

and

$$\rho_{n+1}(v) = a_v.$$

Then,

$$d^2(v) = d(m_v) = 0$$

and

$$\rho_n(dv) = \rho_n(m_v) = da_v = d(\rho_{n+1}(v)).$$

This shows that $\mathfrak{M}(n+1)$ is a differential algebra and that ρ_{n+1} is a map of differential algebras. Lastly, we must show that $H^i(\mathfrak{M}(n+1), \mathcal{A}) = 0$ for $i \leq n+2$. For this we need a lemma.

Lemma 10.4. *(a)* $\mathfrak{M}(n)$ *has no elements of degree* 1.
(b) $H^{n+2}(\mathfrak{M}(n), \mathfrak{M}(n+1)) = V$.
(c) $H^{n+3}(\mathfrak{M}(n), \mathfrak{M}(n+1)) = 0$.

Proof. (a) But for any $n \geq 1$, the DGAs $\mathfrak{M}(n)$ and $\mathfrak{M}(1)$ agree in degrees ≤ 1 so that $\mathfrak{M}(n)$ has no generators of degree ≤ 1.

(b) Let us consider the relative cocycles of degree $(n+2)$. These are all of the form $(a, v+b)$ where $a, b \in \mathfrak{M}(n)$, and $v \in V$. Furthermore, $da = 0$, and $a = dv+db$.

Varying such a cocycle by $d(-b, 0)$ changes it to (a', v). No element of this form is exact unless $v = 0$.

Conversely, given $v \in V$, we have the cocycle (dv, v). This proves that $H^{n+2}(\mathfrak{M}(n), \mathfrak{M}(n+1)) = V$.

(c) Since $\mathfrak{M}(n)$ has no generators of degree ≤ 1, $\mathfrak{M}(n)$ and $\mathfrak{M}(n+1)$ are the same in degree $(n+2)$. It follows easily from this and the fact that $\mathfrak{M}(n) \subset \mathfrak{M}(n+1)$ that $H^{n+3}(\mathfrak{M}(n), \mathfrak{M}(n+1)) = 0$. ■

We have a map of pairs $(\mathrm{id}, \rho_{n+1})\colon (\mathfrak{M}_n, \mathfrak{M}_{n+1}) \to (\mathfrak{M}_n, \mathcal{A})$. The corresponding map of long exact sequences, the above fact, and the five lemma prove that $\rho_{n+1}^*\colon H^*(\mathfrak{M}(n+1)) \to H^*(\mathcal{A})$ is an isomorphism for $* \leq n+1$ and that $\rho_{n+1}^*\colon H^{n+2}(\mathfrak{M}(n+1)) \to H^{n+2}(\mathcal{A})$ is an injection.

This completes the inductive construction of the $(\mathfrak{M}(n)$ and ρ_n. Define $\mathfrak{M} = \bigcup_n \mathfrak{M}(n)$ and define $\rho\colon \mathfrak{M} \to \mathcal{A}$ by $\rho|\mathfrak{M}(n) = \rho_n$. Since cohomology commutes with direct limits, it follows that $\rho^*\colon H^*(\mathfrak{M}) \to H^*(\mathcal{A})$ is an isomorphism. Thus, $(\mathfrak{M}, \rho)$ is a minimal model for $\mathcal{A}$. ■

Chapter 11
Homotopy Theory of DGAs

In this chapter we shall delve more deeply into the homotopy theory of DGAs. One consequence of this study will be to prove the uniqueness of the minimal model.

11.1 Homotopies

Definition. Let f and g be maps of DGA from $\mathcal{A}$ to $\mathcal{B}$. A *homotopy* from f to g is a map

$$\mathrm{H}\colon \mathcal{A} \longrightarrow \mathcal{B} \otimes (\mathrm{t}, \mathrm{dt})$$

satisfying

$$\mathrm{H}|_{\mathrm{t}=0} = \mathrm{f} \quad \text{and} \quad \mathrm{H}|_{\mathrm{t}=1} = \mathrm{g}.$$

Of course, when we restrict to $\mathrm{t} = 0, 1$ we also set $\mathrm{dt} = 0$.

Explanation: $(\mathrm{t}, \mathrm{dt})$ represents the tensor product of polynomials on t (degree of $\mathrm{t} = 0$) with the exterior algebra on dt, where the degree of t is zero and the degree of dt is 1). It is the algebra of polynomial forms on the real line, $\mathbb{R}$. The two restrictions correspond to evaluations of forms at the points $\{0\}$ and $\{1\}$. The idea for this definition comes from dualizing the usual definition on the space level.

To study homotopies, we introduce an additive operator

$$\int_0^1 : \mathcal{B} \otimes (\mathrm{t}, \mathrm{dt}) \longrightarrow \mathcal{B}$$

by

$$\int_0^1 \mathrm{b} \otimes \mathrm{t}^{\mathrm{i}} = 0 \text{ and}$$

P. Griffiths and J. Morgan, *Rational Homotopy Theory and Differential Forms*, Progress in Mathematics 16, DOI 10.1007/978-1-4614-8468-4_11,
© Springer Science+Business Media New York 2013

$$\int_0^1 b \otimes t^i dt = (-1)^{\deg b} \frac{b}{i+1}.$$

Likewise, define

$$\int_0^t : \mathcal{B} \otimes (t, dt) \longrightarrow \mathcal{B} \otimes (t,\ dt)$$

by

$$\int_0^t b \otimes t^i = 0 \text{ and}$$

$$\int_0^t b \otimes t^i\, dt = (-1)^{\deg b}\, b \otimes \frac{t^{i+1}}{i+1}.$$

The following are proved directly from the definitions:

(1) If $\beta \in \mathcal{B} \otimes (t, dt)$, then

$$d\left(\int_0^t \beta\right) + \int_0^t d\beta = \beta - (\beta|_{t=0}) \otimes 1.$$

(2) If $H: \mathcal{A} \rightarrow \mathcal{B} \otimes (t, dt)$ is a homotopy from f to g, then

$$d \int_0^1 H(a) + \int_0^1 dH(a) = g(a) - f(a).$$

(Note that (2) follows from (1) by taking $\beta = H(a)$ and restricting to $t = 1$.)

11.2 Obstruction Theory

The basic result in the homotopy theory of Hirsch extensions is the following:

Proposition 11.1. *Given a diagram*

$$\begin{array}{ccc} \mathcal{A} & \xrightarrow{\ g\ } & \mathcal{B} \\ \downarrow & & \downarrow \varphi \\ \mathcal{A} \otimes_d \Lambda(V^k) & \xrightarrow{\ f\ } & \mathcal{C} \end{array}$$

and a homotopy $H: \mathcal{A} \rightarrow \mathcal{C} \otimes (t, dt)$ *from* $\varphi \circ g$ *to* $f|\mathcal{A}$, *there is an obstruction class* $\mathcal{Q} \in H^{k+1}(\mathcal{B}, \mathcal{C}; V^*)$ *which vanishes if and only if there is an extension* $\tilde{g}: \mathcal{A} \otimes_d \Lambda(V^k) \rightarrow \mathcal{B}$ *of* g *and an extension* $\tilde{H}$ *of* H *to a homotopy from* f *to* $\varphi \circ \tilde{g}$.

Proof. For each $v \in V$, we define $\tilde{\mathcal{Q}}(v) \in \mathcal{B}^{k+1} \oplus \mathcal{C}^k$ by

$$\tilde{\mathcal{Q}}(v) = (g(dv), f(v) + \int_0^1 H\,(dv)).$$

This is the obstruction cocycle for extending g and H. First, we show that $\tilde{\mathcal{Q}}(v)$ is indeed a cocycle, and then we show that if $\tilde{\mathcal{Q}}(v)$ is exact for all $v \in V$, the sought after extensions exist.

$$\begin{aligned} d\tilde{\mathcal{Q}}(v) &= (dg(dv), \varphi \circ g(dv) - df(v) - d\int_0^1 H(dv)) \\ &= (g(d^2v), \tilde{\varphi} \circ g(dv) \ - f(dv) - d\int_0^1 H\,(dv) \ - \int_0^1 dH\,(dv)) \\ &= (0, 0). \end{aligned}$$

Let $[Q]: V \to H^{k+1}(\mathcal{B}, \mathcal{C})$ be the homomorphism induced by $\tilde{Q}$. Such a homomorphism is the same as an element $Q \in H^{k+1}(\mathcal{B}, \ \mathcal{C}; V^*)$.

If $[Q](v) = 0$ for all $v \in V$, then there are relative cochains (b_v, c_v) (depending linearly on v) so that $d(b_v, c_v) = \tilde{Q}(v)$.

Define

$$\begin{aligned} \tilde{g}(v) &= b_v \ \text{ and} \\ \tilde{H}(v) &= f(v) + \int_0^t H(dv) + d(c_v \otimes t). \end{aligned}$$

Let us check that these equations define DGA maps $\tilde{g}: \mathcal{A} \otimes_d \Lambda(V)_k \to \mathcal{B}$ extending g and $\tilde{H}: \mathcal{A} \otimes_d \Lambda(V)_k \to \mathcal{C} \otimes (t, dt)$ extending H. For this, it is necessary only that $d\tilde{g}(v) = g(dv)$ and $d\tilde{H}(v) = H(dv)$. Clearly, $d\tilde{g}(v) = db_v = g(dv)$. Also, by (1), $d\tilde{H}(v) = df(v) + H(dv) - H(dv)|_{t=0}$. Since H is a homotopy from f to $\varphi \circ g$, $H(dv)|_{t=0} = f(dv)$. Thus, $d\tilde{H}(v) = H(dv)$. Lastly, we must show that $\tilde{H}$ is a homotopy from f to $\varphi \circ \tilde{g}$. But

$$\begin{aligned} \tilde{H}(v)|_{t=0} &= f(v) \ \text{ and} \\ \tilde{H}(v)|_{t=1} &= f(v) + \int_0^1 H(dv) + dc_v \\ &= f(v) + \int_0^1 H(dv) + (\varphi(b_v) - f(v) - \int_0^1 H(dv)) \\ &= \varphi(b_v) = \varphi \circ \tilde{g}(v). \end{aligned}$$

Conversely, if we are given any extension $\tilde{g}$ of g and $\tilde{H}$ of H such that $\tilde{H}$ is a homotopy from f to $\varphi \circ \tilde{g}$, then define $\psi_{(\tilde{g},\tilde{H})}: V \to \mathcal{B}^k \oplus \mathcal{C}^{k-1}$ by $\psi(v) = (\tilde{g}(v), \int_0^1 \tilde{H}(v))$. One checks directly that

$$d\psi_{(\tilde{g},\tilde{H})} = (g(dv), f(v) + \int_0^1 H(dv)) = \tilde{\mathcal{Q}}(v).$$

■

We shall also need a relative version of this lifting property.

Lemma 11.2. *Suppose given the following commutative diagram:*

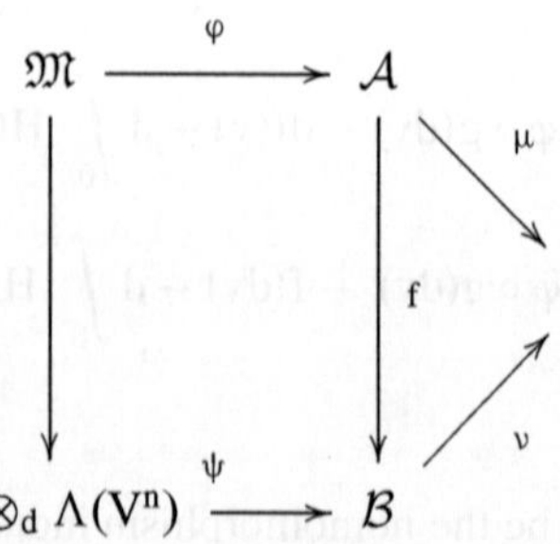

where

(1) $\nu \circ f = \mu$.
(2) μ *is onto.*
(3) $\mu \circ \varphi = \nu \circ \psi|\mathfrak{M}$.
(4) $\mathfrak{M} \xrightarrow{H} \mathcal{B} \otimes (t, dt) \xrightarrow{\nu\otimes 1} \mathcal{C} \otimes (t, dt)$ *is constant (i.e.,* $(\nu \otimes 1) \circ (H) = (\nu \circ H_0) \otimes 1$).

Then, the obstruction cohomology class $[Q] \in H^{n+1}(\mathcal{A}, \mathcal{B}, V^*)$ *vanishes if, and only if, there is an extension* $\tilde{\varphi}$ *of* φ *and an extension* $\tilde{H}\colon \mathfrak{M} \to \mathcal{B} \otimes (t, dt)$ *of* H *satisfying*

(1) $\mu \circ \tilde{\varphi} = \nu \circ \psi$ *and*
(2) $(\nu \otimes 1) \circ \tilde{H}$ *is a constant homotopy.*

Proof. Define $\tilde{Q}\colon V \to \text{Cocycles}^{n+2}(\mathcal{A}, \mathcal{B})$ as before:

$$\tilde{Q}(v) = (\tilde{\varphi}(dv), \varphi(v) + \int_0^1 H(v)).$$

Let $a_v \in \mathcal{A}$ be such that $\mu(a_V) = \nu(\varphi(v))$. Consider $\tilde{Q}(v) - d(a_v, 0) = (\tilde{\varphi}(dv) - da_v, \varphi(v) - a_v + \int_0^1 H(dv))$. This is a cocycle in $(\text{Ker}\,\mu)^{n+1} \oplus (\text{Ker}\,\nu)^n$. If it is exact here, $\tilde{Q}(v) - d(a_v, 0) = d(\alpha_v, \beta_v)$ with $\alpha_v \in \text{Ker}\,\mu$ and $\beta_v \in \text{Ker}\,\nu$.

Then, define $\tilde{\varphi}$ and $\tilde{H}$ using the cochain $(a_V + \alpha_v, \beta_v)$. Checking the formulas in Proposition 11.1, one sees that $\mu \circ \tilde{\varphi} = \nu \circ \psi$ and that $(\nu \otimes 1) \circ \tilde{H}$ is constant. Since μ is onto, the five lemma implies that $\tilde{H}^*(\text{Ker}\mu, \text{Ker}\nu) \to H^*(\mathcal{A}, \mathcal{B})$ is an isomorphism. Thus, $\tilde{Q}(v) - d(a_v, 0)$ is exact in $(\text{Ker}\,\mu, \text{Ker}\,\nu)$ if, and only if, $[\mathcal{Q}(v)] = 0$ in $H^{n+1}(\mathcal{A}, \mathcal{B})$. ■

Corollary 11.3. *Given a commutative diagram*

$$\begin{array}{ccc} \mathfrak{M} & \xrightarrow{\varphi} & \mathcal{A} \\ \downarrow & & \downarrow{\scriptstyle f} \\ \mathfrak{M} \otimes_d \Lambda(V^n) & \xrightarrow{\psi} & \mathcal{B} \end{array}$$

such that f *is onto, the element* $[Q]\colon V \to H^{n+1}(\mathcal{A}, \mathcal{B})$ *is the obstruction to extending* φ *to a map* $\tilde{\varphi}\colon \mathfrak{M} \otimes_d \Lambda(V)^n \to \mathcal{A}$ *such that* $f \circ \tilde{\varphi} = \psi$.

Proof. Apply Lemma 11.2 with $\mathcal{C} = \mathcal{B}$. The cohomology of Ker f is identified with the relative cohomology $H^*(\mathcal{A}, \mathcal{B})$. ■

11.3 Applications of Obstruction Theory

Corollary 11.4. *For* $\mathfrak{M}$ *minimal, the relation on maps from* $\mathfrak{M} \to \mathcal{A}$ *of being homotopic is an equivalence relation.*

Proof. Let $H\colon \mathfrak{M} \to \mathcal{A} \otimes (t_1, dt_1)$ be a homotopy from f_0 to f_1 and $J\colon \mathfrak{M} \to \mathcal{A} \otimes (t_2, dt_2)$ be a homotopy from f_1 to f_2. Let $\mathcal{X}$ be the differential algebra

$$(t_1, t_2, dt_1, dt_2)/\{t_2(t_1 - 1) = 0,\ t_1 dt_2 = t_2 dt_1 = 0\}.$$

This algebra represents the piecewise polynomial forms on the variety $t_2(t_1 - 1) = 0$ in the (t_1, t_2)-plane:

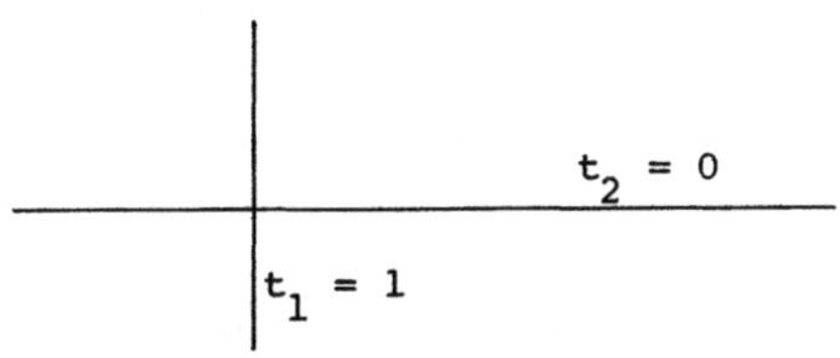

The homotopies H and J define a map "$H + J$"$\colon \mathfrak{M} \to \mathcal{A} \otimes \mathcal{X}$. If

$$H(m) = \Sigma a_i \otimes t_1^i + b_i \otimes t_1^i dt_1$$

and

$$J(m) = \Sigma a_j' \otimes t_2^j + b_j' \otimes t_2^j dt_2,$$

then since $H(m)|_{t_1=1} = J(m)|_{t_2=0}$, we have $\Sigma_{i\geq 0}a_i = a_0'$. The formula for "H + J''(m) is

$$\Sigma a_i \otimes t_1^i + h_i \otimes t^i_{1\,dt} + \Sigma_{j\geq 1}a_j' \otimes t_2^j + \Sigma_{j\geq 0}b_j' \otimes t_2^j dt_2.$$

One checks easily that "H + J" is a map of DGAs. Consider the diagram

$$\begin{array}{ccc} & & \mathcal{A}\otimes(t_1, t_2.dt_1, dt_2) \\ & & \downarrow p \\ \mathfrak{M} & \xrightarrow{\text{"H+J"}} & \mathcal{A}\otimes\mathcal{X} \end{array}$$

The obstructions to lifting "H + J'' lie in $H^*(\mathcal{A}\otimes[(t_1, t_2, dt_1, dt_2)], \mathcal{X}) = 0$.

Since p is onto, Lemma 11.2 says that there is a map $\rho\colon \mathfrak{M} \to \mathcal{A}\otimes(t_1, t_2, dt_1, dt_2)$ such that $p\circ\rho =$ "H + J''. If we restrict ρ to $t_1 = t_2$, we find a homotopy from $f_0 = H|_{t_1=0}$ to $f_2 = J|_{t_2=1}$.

This proves that the relation of being homotopic is transitive. Reflexivity and symmetry are easily shown. ■

Definition. We denote the homotopy classes of maps of DGA from $\mathcal{A}$ to $\mathcal{B}$ by $[\mathcal{A}, \mathcal{B}]$.

Theorem 11.5. *Let $\varphi\colon \mathcal{B} \to \mathcal{C}$ induce an isomorphism on cohomology, and let $\mathfrak{M}$ be a minimal differential algebra. Then $\varphi_*[\mathfrak{M}, \mathcal{B}] \to [\mathfrak{M}, \mathcal{C}]$ is a bijection.*

Proof. If we have $f\colon \mathfrak{M} \to \mathcal{C}$, then the obstructions to lifting f, up to homotopy, to $\mathcal{B}$ step by step over the natural increasing filtration of $\mathfrak{M}$ lie in $H^{n+1}(\mathcal{B}, \mathcal{C}; I^n(\mathfrak{M})^*)$. (Here, $I^n(\mathfrak{M})$ is the space of indecomposables of $\mathfrak{M}$ in degree n. The "$*$" means the dual vector space.) Since $\varphi\colon \mathcal{B} \to \mathcal{C}$ induces an isomorphism on cohomology, $H^*(\mathcal{B}, \mathcal{C}; V) = 0$ for all vector spaces V. Thus, there is a map $g : \mathfrak{M} \to \mathcal{B}$ and a homotopy from f to $\varphi\circ g$. This proves that φ_* is onto.

To show that φ_* is one-to-one, suppose given f_0 and $f_1\colon \mathfrak{M} \to \mathcal{B}$ and a homotopy $H\colon \mathfrak{M} \to \mathcal{C}\otimes(t, dt)$ between $\varphi\circ f_0$ and $\varphi\circ f_1$. Let $\mathcal{P}$ be the kernel of

$$[\mathcal{C}\otimes(t, dt)] \oplus \mathcal{B} \oplus B \xrightarrow{\mu} \mathcal{C} \oplus \mathcal{C} \longrightarrow 0,$$

where the map μ sends $\gamma \in \mathcal{C}\otimes$ (t, dt) to $(\gamma|_{t=0}, \gamma|_{t=1})$ and sends $(b, b') \in \mathcal{B}\oplus\mathcal{B}$ to $(f(\mathcal{B}), f(b'))$. There is a map

$$\rho\colon \mathcal{B}\otimes(t, dt) \longrightarrow \mathcal{P},$$

which sends β to $(f\otimes Id_{(t,dt)}(\beta), \beta|_{t=0}, \beta|_{t=1})$. This map induces an isomorphism on cohomology. To see this, notice that one has a long exact sequence:

$$\longrightarrow H^*(\mathcal{P}) \longrightarrow H^*(\mathcal{C}) \oplus H^*(\mathcal{B}) \oplus H^*(B) \xrightarrow{\mu^*} H^*(\mathcal{C}) \oplus H^*(\mathcal{C}). \longrightarrow$$

The map μ^* sends (c, b_0, b_1) to $(c - f^*b_0, c - f^*b_1)$. Hence, μ^* is onto and $H^*(\mathcal{P}) \subset H^*(\mathcal{C}) \oplus H^*(B) \oplus H^*(\mathcal{B})$ is represented as all triples $(c, (f^*)^{-1}(c), (f^*)^{-1}(c))$. From this, one sees easily that $\rho\colon \mathcal{B} \otimes (t, dt) \to \mathcal{P}$ induces an isomorphism in cohomology.

The maps f_0 and $f_1\colon \mathfrak{M} \to \mathcal{B}$ together with $H\colon \mathfrak{M} \to \mathcal{B} \otimes (t, dt)$ define a map

$$\rho\colon \mathfrak{M} \longrightarrow \mathcal{P}.$$

Apply Lemma 11.2 to the diagram

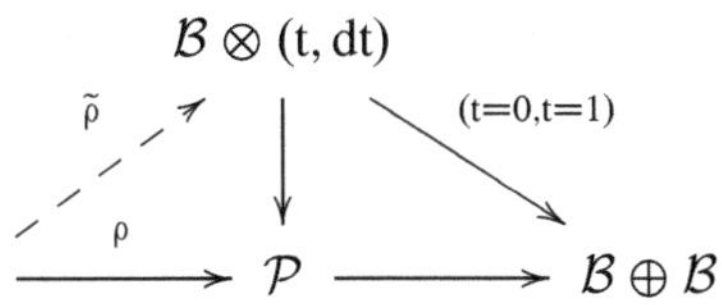

We see that there are no obstructions to lifting ρ to $\tilde{\rho}\colon \mathfrak{M} \to \mathcal{B} \otimes (t, dt)$ such that $\tilde{\rho}|_{i=1} = \pi_{B_i} \circ \rho = f_i$. Thus, f_0 and $f_1\colon \mathfrak{M} \to \mathcal{B}$ are homotopic. ■

11.4 Uniqueness of the Minimal Model

Theorem 11.6. *If $\mathcal{A}$ is a differential algebra and*

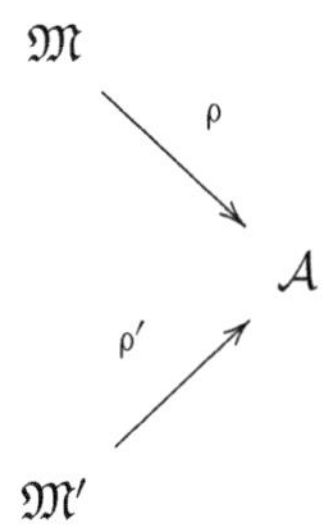

are minimal models for $\mathcal{A}$, then there is an isomorphism $I\colon \mathfrak{M} \to \mathfrak{M}'$ *and a homotopy* H *from* $\rho\colon \rho' \circ I$. *The isomorphism* I *is itself determined by these conditions up to homotopy.*

Proof. Applying Theorem 11.5, we see that there is a map $I\colon \mathfrak{M} \to \mathfrak{M}'$ with $\rho' \circ I$ is homotopic to ρ, and that such an I is well defined up to homotopy. It remains to show that any such $I\colon \mathfrak{M} \to \mathfrak{M}'$ is an isomorphism. Since, on the level of cohomology, $I^* \circ (\rho')^* = \rho^*$ and $(\rho')^*$ and ρ^* are isomorphisms, it follows that any such $I : \mathfrak{M} \to \mathfrak{M}'$ induces an isomorphism on cohomology. To conclude the proof, we have the following lemma:

Lemma 11.7. *Let $\mathfrak{M}$ and $\mathfrak{M}'$ minimal DGAs and suppose that* $I\colon \mathfrak{M} \to \mathfrak{M}'$ *induces an isomorphism on cohomology. Then* I *is an isomorphism of DGAs.*

Proof. Clearly, I induces $I_n : \mathfrak{M}(n) \to \mathfrak{M}'(n)$. Assuming inductively that I_n is an isomorphism, we shall show that I_{n+1} is an isomorphism. Consider the exact sequence of cohomology:

$$\begin{array}{ccccccc}
H^{n+1}(\mathfrak{M}(n)) & \longrightarrow & H^{n+1}(\mathfrak{M}) & \longrightarrow & H^{n+2}(\mathfrak{M}(n), \mathfrak{M}) & \longrightarrow & H^{n+2}(\mathfrak{M}(n)) \\
\downarrow{\scriptstyle I_n^*} & & \downarrow{\scriptstyle I^*} & & \downarrow{\scriptstyle (I_n,I)^*} & & \downarrow{\scriptstyle I_n^*} \\
H^{n+1}(\mathfrak{M}'(n)) & \longrightarrow & H^{n+1}(\mathfrak{M}') & \longrightarrow & H^{n+2}(\mathfrak{M}'(n), \mathfrak{M}') & \longrightarrow & H^{n+2}(\mathfrak{M}'(n))
\end{array}$$

By the inductive hypothesis and the five lemma, now prove that I^* is an isomorphism.

We claim that $H^{n+2}(\mathfrak{M}(n), \mathfrak{M}) = H^{n+2}(\mathfrak{M}(n), \mathfrak{M}(n+1))$. As we have seen before, $H^{n+2}(\mathfrak{M}(n), \mathfrak{M}(n+1)) \cong V^{n+1}$, the vector space of new generators added in going from $\mathfrak{M}(n)$ to $\mathfrak{M}(n+1)$. To prove the claim, note that in degree $\leq n+2$ the relative cochains for $(\mathfrak{M}(n), \mathfrak{M}(n+1))$ and $(\mathfrak{M}(n), \mathfrak{M})$ are equal. In degrees $>$ $n+2$, the former are a subgroup of the latter. The equality of the relative cohomology groups up through degree $(n+2)$ follows immediately from this.

Since these identifications are natural, we have a commutative diagram:

$$\begin{array}{ccc}
H^{n+2}(\mathfrak{M}(n), \mathfrak{M}) & \xrightarrow{(I_n,I)^*} & H^{n+2}(\mathfrak{M}'(n), \mathfrak{M}') \\
\sim\uparrow & & \uparrow\sim \\
V^{n+1}(\mathfrak{M}) & \xrightarrow{\quad I \quad} & V^{n+1}(\mathfrak{M}')
\end{array}$$

Thus, I induces an isomorphism on the vector space of new generators, and hence I_{n+1} induces an isomorphism:

$$\mathfrak{M}(n+1) \xrightarrow{I_{n+1}} \mathfrak{M}'(n+1).$$

■

This completes the proof of Theorem 11.6. ■

Corollary 11.8. *Let $\rho_A\colon \mathfrak{M}_{\mathcal{A}} \to \mathcal{A}$ and $\rho_{\mathcal{B}}\colon \mathfrak{M}_{\mathcal{B}} \to \mathcal{B}$ be minimal models, and let $f\colon \mathcal{A} \to \mathcal{B}$ be a map of differential algebras. There is map $\hat{f}\colon \mathfrak{M}_{\mathcal{A}} \to \mathfrak{M}_{\mathcal{B}}$ and a homotopy from $\rho_{\mathcal{B}} \circ \hat{f} \longrightarrow f \circ \rho_{\mathcal{A}}$. The map $\hat{f}$ is determined up to homotopy by these properties.*

Proof. This is immediate from Theorem 11.5 applied to the following diagram:

$$\begin{array}{ccc} & & \mathfrak{M}_{\mathcal{B}} \\ & \nearrow & \downarrow \rho_{\mathcal{B}} \\ \mathfrak{M}_{\mathcal{A}} & \xrightarrow{f \circ \rho_{\mathcal{A}}} & \mathcal{B} \end{array}$$

since $\rho_{\mathcal{B}}$ induces an isomorphism on cohomology. ■

Example. One note of caution is necessary here. On the object level, a minimal model is uniquely determined. This is <u>not</u> true on the map level. Here is an example of a nonzero map $\mathfrak{N} \xrightarrow{\varphi} \mathfrak{M}$ between minimal DGAs defined over $\mathbb{Q}$ which is homotopic to a constant. Let

$$\mathfrak{N} = \{\alpha^2, \beta^3, \gamma^4 : d\alpha = 0 = d\beta,\ d\gamma = \alpha \wedge \beta\}$$

$$\mathfrak{M} = \{\omega^2, \eta^3 := d\eta = \omega \wedge \omega\}$$

$$\rho(\alpha) = \rho(\beta) = 0, \rho(\gamma) = \omega \wedge \omega$$

This map is homotopic to zero. One way to see this is to give explicitly the homotopy. Anticipating the connection between DGAs over $\mathbb{Q}$ and rational homotopy theory, one can give a geometric argument. On the level of spaces, we have $X = S^2$ which is represented by $\mathfrak{M}$ and Y given by

$$\begin{array}{ccc} K(\mathbb{Q},4) & \longrightarrow & Y \\ & & \downarrow \pi \\ & & K(\mathbb{Q},2) \times K(\mathbb{Q},3). \end{array}$$

The map ρ is realized by a map $f\colon X \to Y$. By the Whitehead theorem, $\pi \circ f \sim$ constant, so that f is homotopic to a map of X into a fiber. This map $X \xrightarrow{f} K(\mathbb{Q}, 4)$ is given by the class $\rho(\gamma) = \omega \wedge \omega = 0$ in $H^4(X, \mathbb{Q})$, and so $f \sim$ constant.

Chapter 12
DGAs and Rational Homotopy Theory

The first seven chapters dealt with classical homotopy theory culminating in the study of the rational Postnikov tower of a (simply connected) space. Chapter 9 was devoted to defining a suitable algebra of differential forms on a simplicial complex. In Chaps. 10 and 11, we studied the homotopy theory of DGAs in general. In this chapter, we shall bring together these two theories and explain how the rational homotopy theory of a simplicial complex is determined by the minimal model of its p.l. forms. In fact, we shall see that $\mathbb{Q}$-Postnikov towers and $\mathbb{Q}$ minimal models are perfectly dual.

12.1 Transgression in the Serre Spectral Sequence and the Duality

The basis for the duality alluded to above is the duality between principal fibrations and Hirsch extensions. To understand this, we study the transgression in the Serre spectral sequence. Let $p: E \to B$ be a principal fibration with fiber $K(\pi, n)$. Let $\{E_r^{p,q}, d_r\}$ be the cohomology Serre spectral sequence of this fibration with coefficients π. Then, $E_2^{p,q} = 0$ for $0 < q < n$ and $E_2^{0,n} = H^n(K(\pi, n), \pi) = \mathrm{Hom}(\pi, \pi)$. As a result $E_r^{0,n} = E_2^{0,n}$ for $r \leq n$. The first nonzero differential on $E_*^{0,n}$ is $d_{n+1}: E_n^{0,n} \to E_n^{n+1,0} = H^{n+1}(B; \pi)$. This map $d_{n+1}: \mathrm{Hom}(\pi, \pi) \to H^{n+1}(B; \pi)$ is the transgression map. The class $d_{n+1}(\iota) \in H^{n+1}(B, \pi)$ is the k-invariant of the fibration. (Here, ι is the class corresponding to the identity homomorphism of π to π.) If we take $\mathbb{Q}$-coefficients instead of π coefficients, then $d_{n+1}: \mathrm{Hom}(\pi, \mathbb{Q}) \to H^{n+1}(B; \mathbb{Q})$. This map is dual to an element $[d_{n+1}] \in H^{n+1}(B; \pi \otimes \mathbb{Q})$. If π itself is a rational vector space (of finite dimension), then $[d_{n+1}] \in H^{n+1}(B; \pi \otimes \mathbb{Q}) = H^{n+1}(B; \pi)$ is again the k-invariant of the fibration. Thus, a principal fibration where the fiber is a local Eilenberg–MacLane space (and the group is of finite dimension over $\mathbb{Q}$) is completely determined by the homotopy group π and $[d_{n+1}] \in H^{n+1}(B; \pi)$.

P. Griffiths and J. Morgan, *Rational Homotopy Theory and Differential Forms*, Progress in Mathematics 16, DOI 10.1007/978-1-4614-8468-4_12,
© Springer Science+Business Media New York 2013

12.2 Hirsch Extensions and Principal Fibrations

On the other hand, given a $\mathbb{Q}$-vector space π of finite dimension and an element $[d] \in H^{n+1}(B;\pi)$, one can construct a Hirsch extension

$$\mathcal{A}^*(B) \otimes_d \Lambda\left((\pi^*)^n\right)$$

$d\colon \pi^* \to \mathcal{A}^{n+1}(B)$ is a map whose image is contained in the closed forms and induces $[d] \in H^{n+1}(B;\pi)$. By the discussion following Lemma 9.3, the Hirsch extensions associated with different maps to closed forms representing the same homomorphism to cohomology produce isomorphic Hirsch extensions. Clearly then, there is a bijection correspondence between principal fibrations over B with fiber a local Eilenberg–MacLane space $K(\pi, n)$ and Hirsch extensions of $A^*(B)$ with new generators π^* in degree n.

Let B be a simplicial complex and let $\rho_B\colon \mathfrak{M}_B \to A^*(B)$ be a minimal model for the p.l. forms on B. Let V be a finite-dimensional rational vector space. We have just seen that there is a one-to-one correspondence between equivalence classes of principal $K(V, n)$-fibrations over B and equivalence classes of Hirsch extensions of $\mathfrak{M}_B$ with the vector space of new generators being the dual vector space V^* to V and the new generators being in degree n. The next result shows that this correspondence is more than an abstract one.

Theorem 12.1. *Let* B *and* E *be simplicial complexes and let* $f\colon E \to B$ *be a simplicial map. Let* $\rho_B\colon \mathfrak{M}_B \to A^*(B)$ *be a minimal model for the p.l. forms on* B. *Let* π *be an abelian group with the property that* $V = \pi \otimes \mathbb{Q}$ *is a finite-dimensional rational vector space. We suppose that* f *is a simplicial model for a principal* $K(\pi, n)$*-fibration Let* $k^{n+1} \in H^{n+1}(B;\pi)$ *be the* k*-invariant of this principal fibration. Let* $\mathfrak{M}' = \mathfrak{M}_B \otimes_d (V^*)^n)$ *be a corresponding Hirsch extension, meaning that* ρ^* *identifies* $[d] \in H^{n+1}(\mathfrak{M}_B; V)$ *with*

$$k^{n+1} \otimes 1 \in H^{n+1}(B;\pi) \otimes \mathbb{Q} = H^{n+1}(B;V).$$

Then, there is a map $\rho'\colon \mathfrak{M}' \to A^*(E)$ *with the following properties:*

(i) ρ' *is a minimal model for the p.l. forms on* E.
(ii) The following diagram commutes:

$$\begin{array}{ccc} \mathfrak{M}_B & \xrightarrow{\rho_B} & A^*(B) \\ \downarrow & & \downarrow{\scriptstyle f^*} \\ \mathfrak{M}' & \xrightarrow{\rho'} & A^*(E) \end{array}$$

In this case, we say that the principal $K(V, n)$-fibration over B and the Hirsch extension of the minimal model $\mathfrak{M}_B$ are *dual*.

12.3 Minimal Models and Postnikov Towers

The duality result for principal $K(\pi, n)$-fibrations and Hirsch extensions leads by induction to the duality of the minimal model for p.l. forms $A^*(B)$ and the rational Postnikov tower of B. We will prove this result in Chap. 16.

Corollary 12.2. *Let* B *be a simply connected simplicial complex whose rational homology is a finite-dimensional rational vector space in each degree, and let*

$$\cdots \longrightarrow B_n \xrightarrow{p_n} B_{n-1} \longrightarrow \cdots$$

be a simplicial model for the Postnikov tower of B *and let* $\mathfrak{M}$ *be a minimal model for* $A^*(B)$ *and let* $\mathfrak{M}(n)$ *be the sub DGA generated by elements in degrees* $\leq n$. *Then,* $\mathfrak{M}(n)$ *is a minimal model for* B_n. *In particular, the principal* $K(\pi_n(B), n)$*-fibration*

$$B_n \longrightarrow B_{n-1}$$

has the property that the associated principal fibration with fiber $K(\pi_n(B) \otimes \mathbb{Q}, n)$ *is dual to the Hirsch extension* $\mathfrak{M}(n-1) \to \mathfrak{M}(n)$. *In particular,* $\pi_n(B) \otimes \mathbb{Q}$ *is isomorphic to the dual to the vector space of generators* $I^n(\mathfrak{M})$.

Proof. We turn the process around and build a minimal model dual to the rational form of the Postnikov tower. We build this by induction on n. Since $\mathfrak{M}(2)$ is the polynomial algebra on $H^2(B; \mathbb{Q})$, with $d = 0$, it follows that $\mathfrak{M}(2)$ is a minimal model for the p.l. forms on a simplicial complex homotopy equivalent to B_2. Suppose that for some $n > 2$, by induction we have constructed $\mathfrak{M}(k) \to A^*(B_k)$ for every $\leq k < n$ and maps $\rho_k: \mathfrak{M}(k) \to A^*(B_k)$ such that

1. $\rho_k: \mathfrak{M}(k) \to A^*(B_k)$ is a minimal model.
2. $\mathfrak{M}(k-1) \subset \mathfrak{M}(k)$ and this inclusion is a Hirsch extension with new generators in degree k.
3. $\rho_k | \mathfrak{M}(k-1) = p_k^* \circ \rho_{k-1}$.

■

Then according to Theorem 12.1 letting $\mathfrak{M}(n-1) \subset \mathfrak{M}(n)$ be a Hirsch extension dual to the rational form of the $K(\pi_n(B), n)$-fibration $B_n \to B_{n-1}$, there is a map $\rho_n: \mathfrak{M}(n) \to A^*(B_n)$ satisfying the three properties listed above. Since this Hirsch extension is dual to the principal fibration, the vector space of new generators is dual to $\pi_n(B_n) \otimes \mathbb{Q} = \pi_n(B) \otimes \mathbb{Q}$. This completes the inductive construction. We set $\mathfrak{M} = \cup_{n=2}^{\infty} \mathfrak{M}(n)$. To complete the proof of the corollary, we need to show that $\mathfrak{M}$ is the minimal model for $A^*(B)$.

Since

$$\cdots \longrightarrow B_n \xrightarrow{p_n} B_{n-1} \longrightarrow \cdots$$

is a simplicial model for the Postnikov tower of B. Fix n. There is a map $f_n\colon B \to B_n$ which, after a rational subdivision of B can assume to be simplicial and which is homotopic to the map from B to the nth-stage of its Postnikov tower. Let K_n be this subdivision. We have the composition

$$f_n \circ \rho_n\colon \mathfrak{M}(n) \to A^*(B_n) \to A^*(K_n).$$

This is an n-minimal model in the sense that the relative cohomology vanishes in degrees$\leq n + 1$. Of course, there is the inclusion$A^*(B) \subset A^*(K_n)$, and the relative homology groups of this pair are trivial. Hence, by Proposition 11.1, there is a map $\mu_n\colon \mathfrak{M}(n) \to A^*(B)$ which when composed with the inclusion into $A^*(K_n)$ is homotopic to $f_n \circ \rho_n$. The map μ_n is also an n-minimal model.

Now, we extend to $\mathfrak{M}(n+1)$. There is a rational subdivision K_{n+1} of K_n and a simplicial map $f_{n+1}\colon K_{n+1} \to B_{n+1}$ which is homotopic to the map from B to the $(n+1)$st-stage of its Postnikov tower. In particular, the following diagram commutes up to homotopy:

$$\begin{array}{ccc}
 & A^*(B) & \\
 {}^{s_n}\swarrow & & \searrow^{s_{n+1}} \\
A^*(K_n) & \xrightarrow{r_{n+1}} & A^*(K_{n+1}) \\
\uparrow & & \uparrow \\
A^*(B_n) & \longrightarrow & A^*(B_{n+1}) \\
\rho_n \uparrow & & \uparrow \rho_{n+1} \\
\mathfrak{M}(n) & \hookrightarrow & \mathfrak{M}(n+1),
\end{array}$$

where s_n, s_{n+1}, and r_{n+1} are the maps induced by the various rational subdivisions. We have $r_{n+1} \circ s_n = s_{n+1}$. Now, deform $f_{n+1} \circ \rho_{n+1}$ by a homotopy to an $(n+1)$-minimal model $\mu_{n+1}\colon \mathfrak{M}(n+1) \to A^*(B)$. It follows that $\mu_{n+1}|\mathfrak{M}(n)$ is homotopic to μ_n. Using Proposition 11.1, we can deform μ_{n+1} by a homotopy until its restriction to $\mathfrak{M}(n)$ agrees with μ_n. In this way, we construct a tower of maps $\mu_k\colon \mathfrak{M}(k) \to A^*(B)$ so that for over every $k' < k$ we have $\mu_k|\mathfrak{M}(k') = \mu_{k'}$. Also, each μ_k is a k-minimal model. Then, the collection of these μ_k define a map $\mu\colon \mathfrak{M} \to A^*(B)$ that is a k-minimal model for every k; that is to say, $\mathfrak{M}$ is the minimal model for $A^*(B)$.

Corollary 12.3. *Let* B *be a simply connected simplicial complex whose rational homology groups are finite-dimensional rational vector spaces in each degree, and let* $\mathfrak{M}$ *be a minimal model for the p.l. forms* $A^*(B)$. *Then:*

1. $I^*(\mathfrak{M})$ *and* $\pi_*(B \otimes \mathbb{Q})$ *are dual rational vector spaces.*
2. $H^*(\mathfrak{M}(n))$ *is identified with* $H^*(B_n; \mathbb{Q})$.

3. *The rational* k*-invariant* $(k^{n+1} \otimes 1) \in H^{n+1}(B_{n-1}; \pi_n(B) \otimes \mathbb{Q})$ *is identified with the class of the Hirsch extension* $\mathfrak{M}(n-1) \subset \mathfrak{M}(n)$*; that is to say, the fibration* $B_n \to B_{n-1}$ *is dual to the Hirsch extension* $\mathfrak{M}(n-1) \subset \mathfrak{M}(n)$.

It follows from this corollary that we can build the rational Postnikov tower of B from the minimal model of the p.l. forms $A^*(B)$ and vice-versa. This means that all the rational homotopy theoretic information about a space B is determined by the minimal model for the p.l. forms on B.

12.4 The Minimal Model of the deRham Complex

The result connecting the minimal model for the p.l. forms on a simplicial complex with the Postnikov tower of the complex has an analogue of the minimal model of the smooth forms on a smooth manifold.

Corollary 12.4. *The minimal model of the* C^∞*, forms on a simply connected* C^∞ *manifold* M, $\mathfrak{M}_\infty(M)$*, is isomorphic to* $\mathfrak{M}_M \otimes_{\mathbb{Q}} \mathbb{R}$ *where* $\mathfrak{M}_M$ *is dual in the sense of Theorem 12.1 to the rational Postnikov tower of* M.

Proof. Choose a C^∞-triangulation of M. This gives a diagram:

$$A^*(M) \otimes_{\mathbb{Q}} \mathbb{R} \longrightarrow A^*_{p.C^\infty}(M) \longleftarrow A^*_{C^\infty}(M).$$

where both inclusions induce isomorphisms on cohomology. Thus, the minimal models of $A^*_{C^\infty}(M)$ and $A_{p.C^\infty *}(M)$ are isomorphic, and the minimal model of $A^*(M)$ tensored with $\mathbb{R}$ is isomorphic to that of $A^*_{p.C^\infty}(M)$. Hence, we have an isomorphism, well defined up to homotopy

$$\mathfrak{M}(A^*(M)) \otimes_{\mathbb{Q}} \mathbb{R} \cong \mathfrak{M}(A^*_{C^\infty}(M)).$$

The corollary now follows immediately from Corollary 12.2. ∎

Corollary 12.5. *Let* Y *be a smooth manifold and* $D_1, \ldots, D_k \subset Y$ *smooth submanifolds which intersect transversally. Let* $D = \cup_{i=1}^k D_i$*, and let* $A^*_{p.C^\infty}(D)$ *be the DGA constructed in Sect. 9.6 of Chap. 9. Suppose that* D *is simply connected. The minimal model of* $A^*_{p,C^\infty}(D)$ *is the real form of a rational minimal model which is dual to the rational Postnikov tower of* D.

Proof. Let K be a C^∞-triangulation of D. There is an inclusion map $A^*_{p.C^\infty}(D) \to A^*_{p.C^\infty}(K)$ which gives rise to a commutative diagram of cochain complexes

$$\begin{array}{ccc} A^*_{p.C^\infty}(D) & \longrightarrow & A^*_{p.C^\infty}(K) \\ & \searrow & \downarrow \\ & & C^*(K;\mathbb{R}) \end{array} .$$

By the result in Sect. 9.6 of Chap. 9, the integration map induces an isomorphism in cohomology $H^*(A^*_{p.C\infty}(D)) \xrightarrow{\sim} H^*(K;\mathbb{R})$. By the piecewise C^∞ deRham theorem, integration induces an isomorphism $H^*(A^*_{p.C\infty}(D)) \xrightarrow{\sim} H^*(K;\mathbb{R})$. Thus, the inclusion $A^*_{p.C\infty}(D) A^*_{p.C\infty}(K)$ is a homotopy equivalence of DGAs. The corollary now follows from Corollary 12.2. ■

Chapter 13
The Fundamental Group

So far we have been considering simply connected spaces. In this chapter, we broaden the definition to include information about the fundamental group. We'find that the theory developed in the previous chapter has a natural analogue which relates DGA of p.l. forms to the universal nilpotent quotient of the fundamental group.

13.1 1-Minimal Models

Let $\mathcal{A}$ be a connected, but not necessarily simply connected, DGA. A 1-*minimal model* for $\mathcal{A}$ is DGA $\mathfrak{M}(1)$ and a map

$$\rho : \mathfrak{M}(1) \longrightarrow \mathcal{A}$$

inducing an isomorphism on H^1 and an injection on H^2 where $\mathfrak{M}(1)$ is an increasing union of Hirsch extensions of degree 1:

$$\begin{aligned} \text{Ground Field } &= \mathfrak{M}(1,0) \subset \mathfrak{M}(1,1) \subset \mathfrak{M}(1,2) \subset \ldots. \\ &\cup_{n=0}^{\infty} \mathfrak{M}(1,n) = \mathfrak{M}(1). \end{aligned}$$

Theorem 13.1. *Let $\mathcal{A}$ be a connected DGA. Then, $\mathcal{A}$ has a* 1*-minimal model. Given two such:*

$$\begin{array}{ccc} \mathfrak{M}(1) & & \\ & \searrow^{\rho} & \\ & & \mathcal{A} \\ & \nearrow_{\rho'} & \\ \mathfrak{M}(1)' & & \end{array}$$

P. Griffiths and J. Morgan, *Rational Homotopy Theory and Differential Forms*,
Progress in Mathematics 16, DOI 10.1007/978-1-4614-8468-4_13,

© Springer Science+Business Media New York 2013

there is an isomorphism $I: \mathfrak{M}(1) \to \mathfrak{M}(1)'$ *and a homotopy* H *from* ρ *to* $\rho' \circ I$.

Proof. Let $\mathfrak{M}(1,1) = \Lambda(H^1(\mathcal{A}))$ and let $\rho_1: \mathfrak{M}(1,1) \to \mathcal{A}$ be defined by sending each cohomology class to a closed form representing it. Given $\rho_n: \mathfrak{M}(1,n) \to A$ with ρ_n inducing an isomorphism on H^1, define V^1_{n+1} to be the kernel of $\rho_n: H^2(M(1,n) \to H^2(\mathcal{A})$. (This is also $H^2(\mathfrak{M}(1,n), \mathcal{A})$.) We choose a splitting $s: H^2(\mathfrak{M}(1,n), \mathcal{A}) \to \text{Cocycles}^2(\mathfrak{M}(1,n), \mathcal{A})$ for the natural quotient map $\text{Cocycles}^2(\mathfrak{M}(1,n), \mathcal{A}) \to H^2(\mathfrak{M}(1,n), \mathcal{A})$, We write $s(v) = (m_v, a_v)$. Define $d: V^1_{n+1} \to \mathfrak{M}(1,n)$ by $d(v) = m_v$; define $\rho: V^1_{n+1} \to \mathcal{A}^1$ by $\rho(v) = a_v$ As Before, one sees that these formulae lead to $\mathfrak{M}(1, n+1) = \mathfrak{M}(1,n) \otimes_d \Lambda(V^1_{n+1})$ and $\rho_{n+1}: \mathfrak{M}(1, n+1) \to \mathcal{A}$ extending ρ_n. One sees that the kernel of $\rho_n^*: H^2(\mathfrak{M}(1,n) \to H^2(\mathcal{A})$ is contained in the kernel of $\iota_n^*: H^2(\mathfrak{M}(1,n) \to H^2(\mathfrak{M}(1, n+1))$ where ι_n is the inclusion $\mathfrak{M}(1,n) \subset \mathfrak{M}(1, n+1)$. Thus, in creating $\mathfrak{M}(1, n+1)$, we have killed the kernel of ρ_n on H^2. But in doing so, we may have created a new nonzero kernel in H^2 at the next stage. We keep repeating the process until we arrive at a stage N with $\rho_N^*: H^2(\mathfrak{M}(1,N) \to H^2(\mathcal{A})$ with trivial kernel. If this never happens, then we construct $\mathfrak{M}(1,n)$ for all $1 \le n < \infty$, and let $\mathfrak{M}(1) = \cup_n \mathfrak{M}(1,n)$ $\rho: \mathfrak{M}(1) \to \mathcal{A}$ being defined by $\rho | \mathfrak{M}(1,n) = \rho_n$ for every n. One sees easily, from the fact that, on $H^2, \operatorname{Ker} \rho_n^* \subset \operatorname{Ker} \iota_n^*$, that $\rho^*: H^2(\mathfrak{M}(1)) \to H^2(\mathcal{A}) = 0$ is an injection. On the other hand, since $\rho_n^*: H^1(\mathfrak{M}(1,n)) \to H^1(\mathcal{A})$ is an isomorphism for all n, $\rho^*: H^1(\mathfrak{M}(1)) \to H^1(\mathcal{A})$ is also an isomorphism.

The proof of the existence of I and a homotopy H is exactly the same as in the simply connected case; see Chap. 11. ■

13.2 $\pi_1 \otimes \mathbb{Q}$

Let π be a finitely presented group. We wish to define "$\pi \otimes \mathbb{Q}$". The method is to replace π by a tower of nilpotent groups, $\{N_i(\pi)\}$, and tensor each of them with $\mathbb{Q}$. Thus, $\pi \otimes \mathbb{Q}$ will be a tower of rational nilpotent groups.

Let $\{\Gamma_i(\pi)\}$ be the terms of the lower central series, i.e., $\Gamma_2(\pi) = [\pi, \pi]$ and inductively $\Gamma_{n+1}(\pi) = [\Gamma_n(\pi), \pi]$. Let $N_i(\pi)$ be $\pi / \Gamma_i(\pi)$. Each $N_i(\pi)$ is a nilpotent group of index of nilpotence i; i.e., all i-fold commutators in $N_i(\pi)$ vanish.

The $N_i(\pi)$ fit together in a tower:

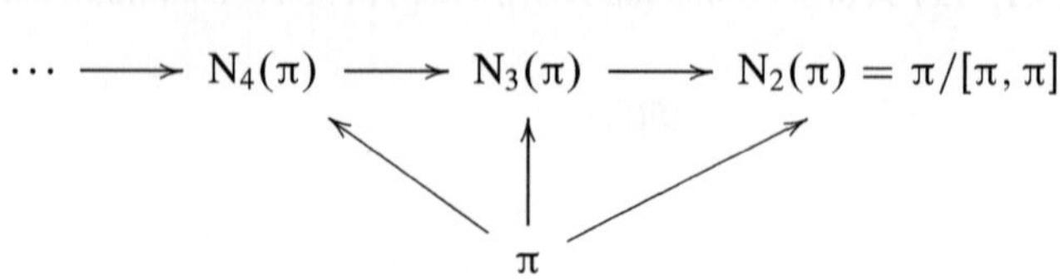

We have short exact sequences (where $\Gamma_n = \Gamma_n(\pi)$)

$$0 \to \Gamma_{n-1}/\Gamma_n \to N_n(\pi) \to N_{n-1}(\pi) \to \{1\}$$

where Γ_{n-1}/Γ_n is an abelian group and is the center of $N_n(\pi)$.

Such extensions correspond to fibrations

$$\begin{array}{ccc} K(\Gamma_{n-1}/\Gamma_n, 1) & \to & K(N_n(\pi), 1) \\ & & \downarrow \\ & & K(N_{n-1}(\pi), 1). \end{array}$$

The fact that Γ_{n-1}/Γ_n is central in $N_n(\pi)$ means that in the fibration, the fundamental group of the base, $N_{n-1}(\pi)$, acts trivially on the fundamental group of the fiber. (In general, the action is conjugation in $N_n(\pi)$.) Hence, central extensions correspond exactly to principal fibrations with base, total space, and fiber having only fundamental groups. The method for tensoring principal fibrations with $\mathbb{Q}$ leads to a method for tensoring nilpotent groups with $\mathbb{Q}$. If inductively, we have

$$K(N_{n-1}(\pi), 1) \to K(N_{n-1}(\pi) \otimes \mathbb{Q}, 1)$$

inducing an isomorphism on rational cohomology, and then we define

$$\begin{array}{ccc} K(\Gamma_{n-1}/\Gamma_n, 1) & \longrightarrow & K((\Gamma_{n-1}/\Gamma_n) \otimes \mathbb{Q}, 1) \\ \downarrow & & \downarrow \\ K(N_n(\pi), 1) & \longrightarrow & K(N_n\pi \otimes \mathbb{Q}, 1) \\ \downarrow & & \downarrow \\ K(N_{n-1}(\pi), 1) & \longrightarrow & K(N_{n-1}\pi \otimes \mathbb{Q}, 1) \end{array}$$

where the fibration on the right has k-invariant in

$$H^2(K(N_{n-1}(\pi) \otimes \mathbb{Q}, 1); \Gamma_{n-1}/\Gamma_n \otimes \mathbb{Q}) = H^2\,(K(N_{n-l}(\pi), 1); \Gamma_{n-1}/\Gamma_n)) \otimes \mathbb{Q}$$

equal to $k \otimes 1_{\mathbb{Q}}$ where k is the k-invariant on the left.

Taking fundamental groups produces a ladder of maps between two central extensions:

$$\begin{array}{ccccccccc} \{0\} & \longrightarrow & \Gamma_{n-1}/\Gamma_n & \longrightarrow & N_n(\pi & \longrightarrow & N_{n-1}(\pi) & \longrightarrow & \{1\} \\ & & \downarrow & & \downarrow \gamma_n & & \downarrow \gamma_{n-1} & & \\ \{0\} & \longrightarrow & \Gamma_{n-1}/\Gamma_n \otimes \mathbb{Q} & \longrightarrow & N_n(\pi) \otimes \mathbb{Q} & \longrightarrow & N_{n-1}(\pi) \otimes \mathbb{Q} & \longrightarrow & \{1\}. \end{array}$$

The extension class for the lower sequence one is just the extension class for the upper one taken with $\mathbb{Q}$-coefficients. Induction on n and a simple comparison theorem of spectral sequences shows that γ_n induces an isomorphism on rational cohomology.

It is a simple argument to show that the elements of finite order in a nilpotent group form a subgroup Tor N. Mal'cev [14] proved that if N is a nilpotent group, then N/Tor N can be embedded in a uniquely divisible nilpotent group $N_{(0)}$. (Uniquely divisible means $x^n = a$ has exactly one solution $x \in N$ for all $a \in N$ and $n \in \mathbb{Z}^+$.) If we take $N_{(0)}$ to be minimal among all uniquely divisible nilpotent groups containing N, then $N_{(0)}$ is determined up to isomorphism by N. It is called the *Mal'cev completion* of N. This is another way to construct the group denoted $N \otimes \mathbb{Q}$ above.

Notice that if the abelianization of π tensored with $\mathbb{Q}$ is a finite-dimensional rational vector space, then for every $n \geq 2$ the rational vector space $\Gamma_{n-1}/\Gamma_n \otimes \mathbb{Q}$ is finite dimensional.

Mal'cev, however, found $N_{(0)}$ by associating to N a rational Lie algebra $\mathscr{L}_N$ and then used the Baker–Campbell–Hausdorff formula, cf, [1], to define a nilpotent group structure, $N_{(0)}$, on $\mathscr{L}_N$. Note that since $\mathscr{L}_N$ is nilpotent, the B–C–H formula becomes a polynomial with rational coefficients and hence defines a group structure on the $\mathbb{Q}$-vector space. This approach has the advantage of proving that the $\mathbb{Q}$-nilpotent groups have $\mathbb{Q}$-nilpotent Lie algebras and hence are the rational points of algebraic groups defined over $\mathbb{Q}$.

We have shown that we can associate to any finitely presented group, π, a tower of rational nilpotent groups, $\{N_n(\pi) \otimes \mathbb{Q}\}$, and tower of rational, nilpotent Lie algebras $\{\mathscr{L}_n(\pi)\}$. These two towers determine each other via the B–C–H formula and its inverse.

Theorem 13.2. *Let* X *be a simplicial complex with* $H_1(X; \mathbb{Q})$ *finite dimensional, and let* $\rho_1 : \mathfrak{M}(1) \longrightarrow A^*(X)$ *be a* 1*-minimal model. Then* $\mathfrak{M}(1)$ *is dual to the tower of Lie algebras* $\{\mathscr{L}_n(\pi_1 X)\}$.

Proof. $\mathfrak{M}_1$ is an increasing union of DGAs $\mathfrak{M}(1,1) \subset \mathfrak{M}(1,2) \subset \ldots$. Let W_n be the generators of $\mathfrak{M}(1,n)$. The differential in $\mathfrak{M}(1,n)$ is determined by $(d|W_n) : W_n \to W_n \wedge W_n$. By definition, the rational Lie algebra dual to $\mathfrak{M}(1,n)$ has underlying vector space $(W_n)^*$. The bracket

$$[\,,\,] : (W_n)^* \wedge (W_n)^* \to W_n^*$$

is dual to d. The Jacobi identity for $[\cdot,\cdot]$ is dual to the equation $(d^2|W_n) = 0$. We call this Lie algebra structure $\mathscr{L}_n$. The inclusion $\mathfrak{M}(1,n) \subset \mathfrak{M}(1,n+1)$ gives an inclusion $W_n \subset W_{n+1}$. Dualizing gives a map of rational Lie algebras:

$$0 \longrightarrow K_{n+1} \longrightarrow W_{n+1}^* \longrightarrow W_n^* \longrightarrow 0.$$

The fact that $(d|W_{n+1}) : W_{n+1} \to W_n \wedge W_n$ dualizes to the fact that K_{n+1} is an abelian Lie algebra and is the center of the Lie algebra $\mathcal{L}_{n+1}$. Since $\mathcal{L}_0$ is the trivial Lie algebra, it follows by induction that each $\mathcal{L}_n$ is a nilpotent Lie algebra.

This gives the duality between 1-minimal models and towers of nilpotent Lie algebras. It remains to show that this duality carries the 1-minimal model for the p.l. forms on X to the tower of Lie algebras associated to its fundamental group. This is first proved inductively for towers of principal fibrations of $K(\pi, 1)$'s. The proof uses the Hirsch lemma and is the complete analogue of Corollary 11.3. Given X, we map it to the following tower:

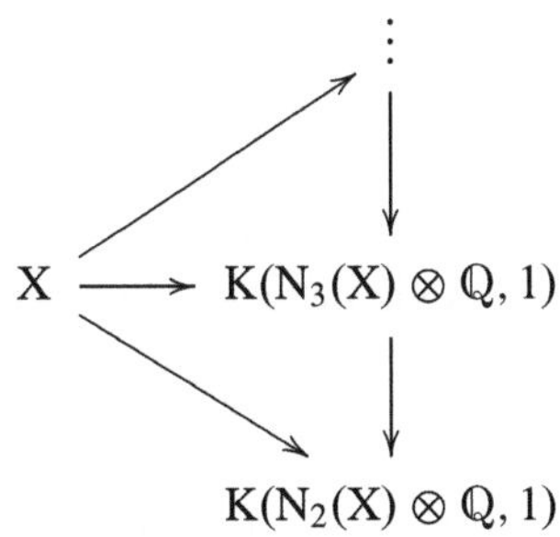

The minimal model for the tower is the one dual to the tower of Lie algebras associated with it. On the other hand,

$$H^1(K(N_i(\pi) \otimes \mathbb{Q}, 1); \mathbb{Q}) = H^1(X; \mathbb{Q})$$

for all i, and

$$\varinjlim\{H^2(K(N_i(\pi) \otimes \mathbb{Q}, 1); \mathbb{Q})\} \longrightarrow H^2(X; \mathbb{Q})$$

is an injection. Thus, the minimal model for this tower pulls back to the 1-minimal model of X. ■

Definition. By the *real fundamental group* of X, we mean the tower of real nilpotent Lie algebras: $\{N_n(\pi_1(X) \otimes \mathbb{R}\}$.

13.3 Functorality

It is when we consider functorality that the base point makes its appearance. We define the base point of a differential algebra to be a map $A^0 \to k$ (k = ground field). If X is a simplicial complex and $p \in X$ is a base point (with rational barycentric coordinates), then it defines $A^0(X) \to \mathbb{Q}$ by evaluation. If $\mathcal{A} \to k$ and $\mathcal{B} \to k$ are differential algebras with base points, then a base point preserving map $\varphi: \mathcal{A} \to \mathcal{B}$ is one that commutes with the maps to k. A homotopy

$$H: \mathcal{A} \to \mathcal{B} \otimes (t, dt)$$

is *base point preserving* if

$$\begin{array}{ccc} \mathcal{A} & \longrightarrow & \mathcal{B} \otimes (\mathrm{t}, \mathrm{dt}) \\ \downarrow & & \downarrow \\ \mathrm{k} & \longrightarrow & \mathrm{k} \otimes (\mathrm{t}, \mathrm{dt}) \end{array}$$

commutes. Notice that if f $\mathcal{B}^0 = \mathrm{k}$, and if H is base point preserving, then $\mathrm{H}(a) = \sum_{i=0}^{\infty} \mathrm{b_i} \otimes \mathrm{t^i}$ for every $a \in \mathcal{A}^1$. A closer look at the argument given in the proof of Theorem 13.1 shows the following:

Theorem 13.3.

(i) *If $\mathcal{A} \to \mathrm{k}$ is a differential algebra with base point, then there is a 1-minimal model $\mathfrak{M}(1)$ which automatically has a unique base point and a base point preserving map*

$$\begin{array}{ccc} \mathfrak{M}(1) & \longrightarrow & \mathcal{A} \\ & \searrow \quad \swarrow & \\ & \mathrm{k} & \end{array}$$

inducing an isomorphism on H^1 and an injection on H^2

(ii) *Given two such $\rho\colon \mathfrak{M}_1 \longrightarrow \mathcal{A}$ and $\rho'\colon \mathfrak{M}(1)' \longrightarrow \mathcal{A}$, then there is an isomorphism $\mathrm{I}\colon \mathfrak{M}(1) \longrightarrow \mathfrak{M}(1)'$ (automatically base point preserving) and a base point preserving homotopy H from $\rho_1' \circ \mathrm{I}$ to ρ. The isomorphism I is well defined up to base point preserving homotopy.*

Completely analogously to 11.2, we have the following:

Theorem 13.4. *If $\mathrm{f}\colon \mathrm{X} \to \mathrm{Y}$ is a base point preserving map, then it induces $\hat{\mathrm{f}}\colon \mathfrak{M}(1)_{\mathrm{Y}} \to \mathfrak{M}(1)_{\mathrm{X}}$ well defined up to base point preserving homotopy.*

At this point, a miraculous thing happens.

Lemma 13.5. *If $\mathrm{H}\colon \mathfrak{M}(1) \to \mathfrak{N}(1) \otimes (\mathrm{t}, \mathrm{dt})$ is a base point preserving homotopy between 1-minimal DGAs, then H is constant in the sense that $\mathrm{H} = \mathrm{H_0} \otimes 1$.*

Proof. We prove by induction that

$$\mathrm{H}|\mathfrak{M}(1, \mathrm{k}) = (\mathrm{H_0}|\mathfrak{M}(1, \mathrm{k})) \otimes 1.$$

Suppose we know this for $\mathrm{k} < \mathrm{n}$, and let x be an element of degree one in $\mathfrak{M}(1, \mathrm{n})$. Then, $\mathrm{H(x)} = \sum_{\mathrm{i}} \eta_{\mathrm{i}} \otimes \mathrm{t^i} + \mathrm{c_i} \otimes \mathrm{t}^i\mathrm{dt}$ for one forms η_i and constants $\mathrm{c_i}$. Since H is base point preserving and the base point map $\mathfrak{N} \to \mathrm{k}$ is an isomorphism in degree zero, the $\mathrm{c_i} = 0$ for all i. Thus, $\mathrm{H(x)} = \sum_i \eta_{\mathrm{i}} \otimes \mathrm{t^i}$, and hence, $\mathrm{H(dx)} = \sum_i \mathrm{d}\eta_{\mathrm{i}} \otimes \mathrm{t^i} - \sum_i \eta_{\mathrm{i}} \otimes \mathrm{t^{i-1}dt}$. But $\mathrm{dx} \in \mathfrak{M}(1, \mathrm{n}-1)$ and hence by induction H(dx) is of the form $\mu \otimes 1$. Consequently, $\eta_i = 0$ for $i > 0$, and hence, $H(x) = \eta_0 \otimes 1$. Since $\mathfrak{M}_{(1,n)}$ is generated by elements in degree 1, it follows that $H|\mathfrak{M}(1,n) = H_0|\mathfrak{M}(1,n) \otimes 1$. Taking the union over all n gives the result. ■

Thus, we have shown that assigning to based simplicial complexes their 1-minimal models is a functor from based homotopy category of simplicial complexes to 1-minimal models and base point preserving maps between them. If we dualize to Lie algebras and then exponentiate, the result is the functor that assigns $\pi_1(X) \otimes \mathbb{Q}$ to X and $f_\#\colon \pi_1(X,p) \otimes \mathbb{Q} \to \pi_1(Y,q) \otimes \mathbb{Q}$ to $f\colon (X,p) \to (Y,q)$. Thus, from the $\mathbb{Q}$-polynomial forms on X, one has a purely algebraic way using p.l. forms on X to recover the homotopy functor "$\pi_1 \otimes \mathbb{Q}$".

13.4 Examples

(1) $S^1 \vee S^1$. Consider

$$0 \longrightarrow A^* \longrightarrow A^*_{C^\infty}(S^1) \oplus A^*_{C^\infty}(S^1) \xrightarrow{\text{ev}} \mathbb{R} \longrightarrow 0.$$

where the map ev is evaluation at 0. The differential algebra A^* can be used to calculate the real fundamental group of $S^1 \vee S^1$. There is a map of differential algebras

$$\{H^*(S^1 \vee S^1; \mathbb{R}),\ d = 0\} \hookrightarrow A^*$$

which induces an isomorphism on cohomology. Hence, to construct the 1-minimal model for A^*, it suffices to construct the 1-minimal model for $H^*(S^1 \vee S^1; \mathbb{R})$. We begin with

$$\Lambda(\{x,y\}) \longrightarrow H^*(S^1 \vee S^1)$$

where x and y are a basis for $H^1(S^1 \vee S^1)$. This map induces an isomorphism on H^1, but it has a kernel in degree 2 generated by $x \wedge y$. The next stage is

$$\rho_2\colon \Lambda(\{x,y,\eta\}) \longrightarrow H^*(S^1 \vee S^1)$$

where $d\eta = x \wedge y$ and $\rho_2(\eta) = 0$. This map has a kernel in degree 2 generated by $x \wedge \eta$ and $y \wedge \eta$. As we continue the construction, the kernels which we encounter in degree 2 keep growing larger and the construction must be repeated *ad infinitum*. Actually, one is constructing the increasing system of 1-minimal DGAs dual to the tower of free nilpotent Lie algebras on 2 generators. Clearly, this is the tower associated to the free group on 2 generators.

In general, when $H^2(X, \mathbb{Q}) = 0$, the tower of Lie algebras constructed is a tower of free nilpotent Lie algebras.

(2) Let N be the nil-manifold obtained by dividing the group of upper triangular real matrices

$$\left\{\begin{pmatrix} 1 & x & z \\ 0 & 1 & y \\ 0 & 0 & 1 \end{pmatrix}\right\}$$

by the lattice of such integral matrices. This gives a compact 3-manifold whose cohomology ring is the same as that of $S^1 \times S^2 \# S^1 \times S^2$. When we build the 1-minimal model for the forms on $S^1 \times S^2 \# S^1 \times S^2$, we get the infinite process described in Example 1. When we build the 1-minimal model for N, we find that the result is

$$\Lambda(\{x, y\}) \otimes_d \Lambda(\eta); \qquad d\eta = x \wedge y.$$

Thus, $H^2(N)$ is generated by the "Massey Products" $\langle x, x, y \rangle = [x \wedge \eta]$ and $\langle x, y, y \rangle = [y \wedge \eta]$. In fact, the 1-minimal model for N is its minimal model.

Chapter 14
Examples and Computations

14.1 Spheres and Projective Spaces

Let us consider an odd sphere S^{2n+1}. The first stage in building the minimal model for the forms on S^{2n+1} is to construct an exterior algebra $\Lambda(e)$ on a generator of degree $(2n + 1)$. Clearly, the cohomology of this DGA maps isomorphically to $H^*(S^{2n+1})$ when we send e to a closed form on S^{2n+1} which is not exact. Thus, $\Lambda(e)$ is the minimal model for the forms on S^{2n+1}. By Corollary 12.4, this implies that $\pi_i(S^{2n+1}) \otimes \mathbb{Q}$ is zero for $i \neq 2n + 1$ and equal to $\mathbb{Q}$ for $i = 2n + 1$.

Let $A^*(S^{2n})$ be the forms on S^{2n}. The first stage in building the minimal model for $A^*(S^{2n})$ is the polynomial algebra, $P(x_{2n})$, on a generator of degree 2n. Clearly, x_{2n}^2 is cohomologous to zero in $A^*(S^{2n})$. Thus, we tensor in an exterior algebra $\Lambda(y_{4n-1})$ with $dy = x^2$ The product $P(x_{2n}) \otimes_d \Lambda(y_{4n-1})$ is the minimal model. It follows that

$$\pi_i(S^{2n}) \otimes \mathbb{Q} \cong \begin{cases} \mathbb{Q} & i = 2n, 4n-1 \\ 0 & i \neq 2n, 4n-1. \end{cases}$$

If $f\colon S^n \to X$ is a simplicial map (with $n > 1$ and X simply connected), then there is induced $\hat{f}\colon \mathfrak{M}_X \to \mathfrak{M}_{S^n}$ well-defined up to homotopy. In degree n we have $I^n(\mathfrak{M}_X)^* \to I^n(\mathfrak{M}_{S^n})^* \cong \mathbb{Q}$. The isomorphism $I^n(\mathfrak{M}_{S^n})^* \cong \mathbb{Q}$ is via integration of the form over the fundamental cycle of S^n. The induced map $I^n(\mathfrak{M}_X)^* \to \mathbb{Q}$ depends only on the homotopy class of $\hat{f}$, and hence only on the homotopy class of f. This defines a map

$$\pi_n(X) \to \mathrm{Hom}_{\mathbb{Q}}(I^n(\mathfrak{M}_X), \mathbb{Q}),$$

and hence a map

$$\pi_n(X) \otimes \mathbb{Q} \to [I^n(\mathfrak{M}_X)]^*.$$

As we shall see in Chap. 15, this is exactly the duality between $I^n(\mathfrak{M}_X)$ and $\pi_n(X) \otimes \mathbb{Q}$.

P. Griffiths and J. Morgan, *Rational Homotopy Theory and Differential Forms*, Progress in Mathematics 16, DOI 10.1007/978-1-4614-8468-4_14,

© Springer Science+Business Media New York 2013

Let us consider the minimal model for the forms on complex projective n-space, $\mathbb{C}P^n$. The first stage of the minimal model is the polynomial algebra on a 2-dimensional generator $P(x_2)$. The ideal generated by the class $(x_2)^{n+1}$ is the kernel of the map $P(x_2) \to H^*(\mathbb{C}P^n; \mathbb{R})$. Thus, the minimal model for the forms on $\mathbb{C}P^n$ is $P(x_2) \otimes_d \Lambda(y_{2n+1}); dy = x^{n+1}$. In particular,

$$\pi_i(\mathbb{C}P^n) \otimes \mathbb{Q} = \begin{cases} 0 & i \neq 2, 2n+1 \\ \mathbb{Q} & i = 2, 2n+1. \end{cases}$$

One can also deduce this from the calculation of the homotopy groups of S^{2n+1} and the fibration $S^1 \to S^{2n+1} \to \mathbb{C}P^n$.

14.2 Graded Lie Algebras

Suppose $\mathscr{L} = \oplus_n \mathscr{L}_n$ is a graded vector space (over a field of characteristic 0). Let $[\,,\,] : \mathscr{L} \otimes \mathscr{L} \to \mathscr{L}$ be a map which is homogeneous of degree 0. We say that $(\mathscr{L}, [\,,\,])$ is a graded Lie algebra if:

(1) $[x, y] = (-1)^{(p+1)(q+1)}[y, x]$ for $x \in \mathscr{L}_p$ and $y \in \mathscr{L}_q$ (graded skew symmetry).
(2) $[x, [y, z]] = [[x, y], z] + (-1)^{pq}[y, [x, z]]$ for $x \in \mathscr{L}_p$ and $y \in \mathscr{L}_q$ (Jacobi identity).

An ordinary Lie algebra L is a graded Lie algebra in which $\mathscr{L}_0 = L$ and $\mathscr{L}_n = 0$ for all $n \neq 0$.

If X is a simply connected space, then the Whitehead product (see Exercise 34) is a bilinear map:

$$\pi_p(X) \otimes \pi_q(X) \xrightarrow{[\,,\,]} \pi_{p+q-1}(X)$$

The formulae in Exercise 34 show that if we define $\mathscr{L}_n = \pi_{n+1}(X) \otimes \mathbb{Q}$, then the Whitehead product makes $\mathscr{L} = \oplus_{n\geq 1}\mathscr{L}_n$ into a graded Lie algebra.

Another source of examples of graded Lie algebras is DGAs. Let $\mathfrak{M}$ be a DGA that is free as a graded commutative algebra on positive dimensional generators. Let $I(\mathfrak{M}) = \mathfrak{M}^+/\mathfrak{M}^+ \wedge \mathfrak{M}^+$ where $\mathfrak{M}^+$ is the ideal of elements of positive degree. Denote by $(\mathfrak{M}^+)^k$ the kth-power of this ideal; i.e.,

$$(\mathfrak{M}^+)^k = \{\omega | \omega = \Sigma_i \alpha_{i1} \wedge \ldots \wedge \alpha_{ik} \text{ with } \alpha_{ij} \in \mathfrak{M}^+\}.$$

If $d\colon \mathfrak{M} \to \mathfrak{M}$ is decomposable, then $d\colon (\mathfrak{M}^+)^k \to (\mathfrak{M}^+)^{k+1}$. Hence, it induces a map

$$\begin{array}{ccc} \mathfrak{M}^+/(\mathfrak{M}^+)^2 & \longrightarrow & (\mathfrak{M}^+)^2/(\mathfrak{M}^+)^3 \\ \downarrow = & & \downarrow = \\ d : I(\mathfrak{M}) & \longrightarrow & I(\mathfrak{M}) \wedge I(\mathfrak{M}) \end{array}$$

The fact that $d^2 = 0$ implies that the composition

$$I(\mathfrak{M}) \xrightarrow{d} I(\mathfrak{M}) \xrightarrow{d \wedge id + (-1)^{\deg} id \wedge d} I(\mathfrak{M}) \wedge I(\mathfrak{M}) \wedge I(\mathfrak{M}) \tag{14.1}$$

is zero. Let $\mathscr{L}_n = [I^{n+1}(\mathfrak{M})]^*$. Dual to d is a map $[\ ,\ ]: \mathscr{L} \otimes \mathscr{L} \to \mathscr{L}$ which is homogeneous of degree 0. The fact that d maps into $I(\mathfrak{M}) \wedge I(\mathfrak{M})$ dualizes to the fact that $[,]$ satisfies the symmetry condition to be a graded Lie algebra bracket. The dual to Eq. (14.1) is the Jacobi identity.

There is a connection between these two examples. If X is a simply connected space, and if $\mathfrak{M}_X$ is its minimal model, then the graded Lie algebras $(\pi_{*+1}(X) \otimes \mathbb{Q}$, Whitehead product) and $(\oplus I^{n+1}(\mathfrak{M}_X)^*, d^*)$ are isomorphic under the map:

$$\pi_{n+1}(X) \otimes \mathbb{Q} \xrightarrow{\sim} [I^{n+1}(\mathfrak{M}_X)]^*.$$

As an example of this, let us consider $S^2 \vee S^2$. Its minimal model begins with two generators ξ_1 and ξ_2 of degree 2. We then add 3-dimensional generators to kill all 4-dimensional cohomology, and so on. The dual graded Lie algebra to this DGA is the free graded Lie algebra on two 1-dimensional generators ξ_1^* and ξ_2^*.

More generally, $S^{n_1} \vee \ldots \vee S^{n_k}$ has a minimal model whose indecomposables are dual to a free graded Lie algebra on generators of degrees $(n_1 - 1), \ldots, (n_k - 1)$ (assuming each $n_i \geq 2$).

14.3 The Borromean Rings

There is a geometric form of Poincaré duality. It allows one to do the calculations involved in building the minimal model with submanifolds instead of forms. The basis for the duality is the following. Let $M^k \subset N^n$ be an embedded submanifold. Suppose that M^k and N^n are both closed and oriented. By the tubular neighborhood theorem there is a neighborhood $\nu(M \subset N)$ which is diffeomorphic to a disk bundle over M:

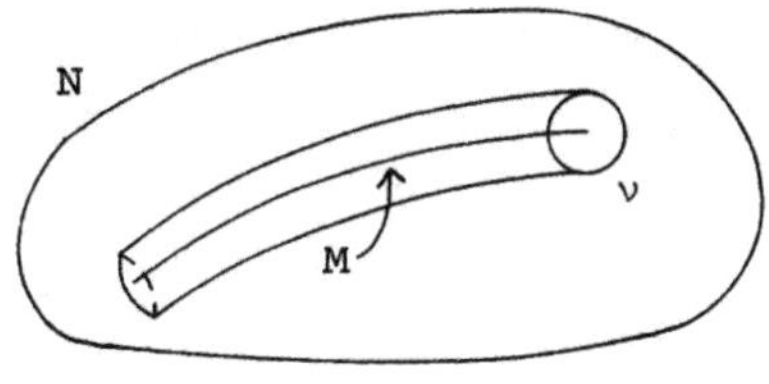

Let D_0^{n-k} be the fiber over a point $m_0 \in M$. Since M and N are oriented, D_0^{n-k} receives an orientation. By the Thom isomorphism theorem (see the exercises), there is a unique class $U_M \in H^{n-k}(\nu(M \subset N), \partial\nu(M \subset N); \mathbb{Z})$ so that $\int_{D_0} U_M = 1$. (Here we assume that M is connected.) There is a C^∞ differential form representing U which vanishes identically near $\partial\nu(M \subset N)$. If we extend by 0 to $N - \nu(M \subset N)$, then we have a closed C^∞-form $\tilde{U}_M$ on all of N. Its cohomology class is the Poincaré dual of $[M] \in H_k(N)$.

There is a form of this duality for manifolds with boundary. If $(M, \partial M) \subset (N, \partial N)$ with M meeting ∂N normally in ∂M, then the same construction yields a class $\tilde{U}_M \in H^{n-k}(N)$ which is the Lefschetz dual of the class $[M, \partial M] \in H_k(N, \partial N)$.

Under the correspondence $M \to \tilde{U}_M$, transverse intersection of manifolds corresponds to wedge product of forms. Thus, if M_0^k and M_1^ℓ are transverse in N^n with intersection $M_{0,1}^{k+\ell-n}$, and if $\tilde{U}_0$ and $\tilde{U}_1$ are Thom forms for M_0 and M_1 supported in sufficiently small tubes, then $\tilde{U}_0 \wedge \tilde{U}_1$ is a Thom form for $M_{0,1}$. This means that $\tilde{U}_0 \wedge \tilde{U}_1$ is a closed form supported in a tube about $M_{0,1}$ and integrating to 1 over each fiber.

The operation of finding a solution for $d\eta = \tilde{U}_M$, given M and $\tilde{U}_M$, corresponds to finding a submanifold of N whose boundary is M. Thus, if $M^k = \partial L^{k+1}$, then there is a form $\tilde{U}_L$ supported in a tube about L, closed outside the tube about M, integrating to 1 over fibers of $\nu(L \subset N)$ which are outside the tube about M, and so that $d\tilde{U}_L = \tilde{U}_M$

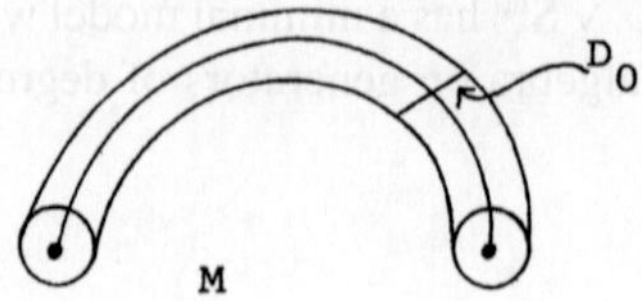

$$\int_{D_0} \tilde{U}_L = 1, \quad d\tilde{U}_L = \tilde{U}_M.$$

We consider now an explicit example of the operation of building the minimal model via submanifolds, namely, the ambient space in S^3-(Borromean rings).

$\mathfrak{B}$ = 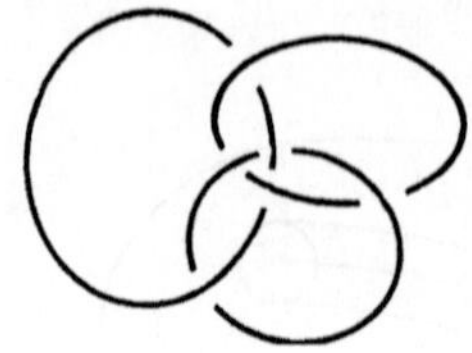 (Borromean rings)

We can think of this as a manifold with boundary by taking out open solid tori around each of the circles, or more simply we work with ordinary cohomology and proper submanifolds

The first cohomology of $S^3 - \mathfrak{B}$ has rank 3 and is generated by classes which are dual to 2-disks spanning the components. We choose these disks as pictured below:

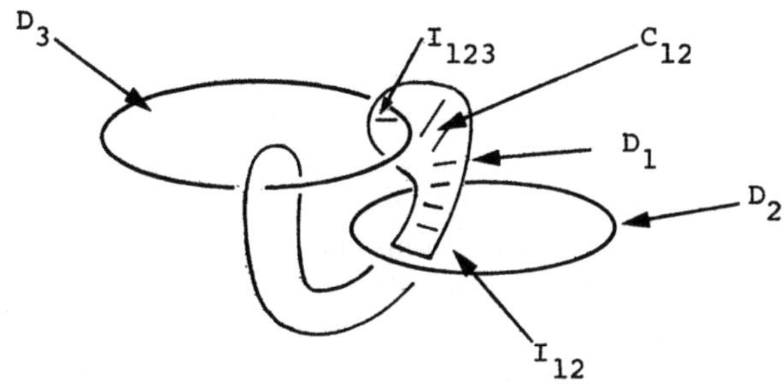

Let $\tilde{U}_1, \tilde{U}_2$, and $\tilde{U}_3$ be the dual Thom classes in $H^1(S^3 - \mathfrak{B})$. The first stage in the minimal model is $\Lambda(\tilde{U}_1, \tilde{U}_2, \tilde{U}_3)$.

The geometric fact that the linking number of any pair is zero in $S^3 - \mathfrak{B}$ means that $\tilde{U}_i \cup \tilde{U}_j = 0$ in $H^2(S^3 - \mathfrak{B})$ for all $i \neq j$. Clearly, $\tilde{U}_2 \wedge \tilde{U}_3 = 0$ as a form since $D_2 \cap D_3 = \emptyset$. The form $\tilde{U}_1 \wedge \tilde{U}_2$ is the Thom form for the interval I_{12}. To solve the equation $d\xi = \tilde{U}_1 \wedge \tilde{U}_2$, we must find a proper 2-dimensional submanifold whose boundary is I_{12}. We take this to be the part of D_1 cut off by I_{12} which lies above D_2, denoted C_{12} in the above drawing. To form the Massey product (see Sect. 14.9 for a definition of Massey products) $< \tilde{U}_1, \tilde{U}_2, \tilde{U}_3 >$, we must take $\eta_{12} \wedge \tilde{U}_3 + \tilde{U}_1 \wedge \eta_{23}$ where $d\eta_{ij} = \tilde{U}_i \wedge \tilde{U}_j$. In this case, we take $\eta_{23} = 0$ and we take η_{12} to be supported near C_{12}. Thus, $\eta_{12} \wedge \tilde{U}_3 + \tilde{U}_1 \wedge \eta_{23}$ is represented by the Thom form of $C_{12} \cap D_3$. This intersection is I_{123}. Since I_{123} is an arc with end points on different components of $\mathfrak{B}$, the class $[I_{123}] \in H_1(S^3 - \mathfrak{B})$ is nonzero. Thus, $< \tilde{U}_1, \tilde{U}_2, \tilde{U}_3 > \neq 0$. If we do a similar calculation for $S^3 - \mathfrak{B}'$ where $\mathfrak{B}'$ is three unlinked circles, then all Massey products $< \tilde{U}_{i_1}, \tilde{U}_{i_2}, \tilde{U}_{i_3} >$ are trivial. Thus, the third stages of the minimal models for the forms on $S^3 - \mathfrak{B}$ and $S^3 - \mathfrak{B}'$ are different. In particular, $\mathfrak{B}$ and $\mathfrak{B}'$ are not isotopic. In fact, since the minimal models differ at the third stage, this implies that $\pi_1(S^3 - \mathfrak{B})/\Gamma_5$ and $\pi_1(S^3 - \mathfrak{B}')/\Gamma_5$ are not isomorphic groups. The group $\pi_1(S^3 - \mathfrak{B}')$ is free. The existence of nonzero Massey products in $S^3 - \mathfrak{B}$ means that its fundamental group is not free, and indeed its 5th nilpotent quotient is different from that of the free group on 3 generators.

14.4 Symmetric Spaces and Formality

A space is said to be *formal* if the homotopy type of the DGA of forms on the space is the same as the homotopy type of the cohomology ring of the space. Thus, if X is formal and $\mathfrak{M}_X$ is a minimal model for the forms on X, then there is a DGA map $\mathfrak{M}_X \to (H^*(X), d = 0)$ which induces the identity on cohomology.

One can always define a map of cochain complexes

$$(H^*(X), d = 0) \to A^*(X)$$

which induces the identity on cohomology by choosing (linearly) closed form representatives for each cohomology class. Usually, this map will not be multiplicative. If it is possible to choose this map to be multiplicative, then $(H^*(X), d = 0)$ and $A^*(X)$ have the same minimal model. Thus, this gives algebraic topological conditions on a space which must be satisfied if there is to be a multiplicative mapping from the cohomology to the forms inducing the identity on cohomology.

If X is a closed Riemannian manifold, then there is a canonical map $(H^*(X), d = 0) \to A^*(X)$ which assigns to each cohomology class its unique harmonic representative. From the above discussion, we see that for X to admit a Riemannian metric in which the wedge product of harmonic forms is harmonic, it must be the case that X is formal (over $\mathbb{R}$).

There is one class of Riemannian manifolds in which wedge product of harmonic forms is harmonic. These are the Riemannian locally symmetric spaces.

14.5 The Third Homotopy Group of a Simply Connected Space

Let A^* be a DGA with $H^1(A^*) = 0$. The first stage in constructing the minimal model of A^* is the polynomial algebra on $H^2 = H^2(A^*)$. We denote this DGA by $\{P[H^2], d = 0\}$. The next stage is created by tensoring in an exterior algebra on the relative 4th-cohomology group $H^4(P[H^2], A^*)$. If X is a simply connected space, then $\pi_3(X) \otimes \mathbb{Q}$ is dual to $H^4(P[H^2], A^*(X))$.

The long exact sequence of the pair $(P[H^2], A^*(X))$ leads to a long exact sequence:

$$0 \longrightarrow H^3(X) \xrightarrow{h^*} \mathrm{Hom}(\pi_3(X), \mathbb{Q}) \longrightarrow P_2[H^2(X)] \xrightarrow{\cup} H^4(X)$$

where h^* is dual to the Hurewicz homomorphism, $P_2[H^2(X)]$ is the vector space of quadratic polynomials in $H^2(X)$, and the map to $H^4(X)$ is the cup product mapping.

If $f: S^3 \to X$ is a continuous mapping, then we can deform it until it becomes simplicial. At this point, it induces a map $f^*: A^*(X) \to A^*(S^3)$, and hence,

$$\hat{f}: \mathfrak{M}_X \longrightarrow \mathfrak{M}_{S^3} = \Lambda(e).$$

There is the evaluation map $\mathfrak{M}_{S^3} \longrightarrow \mathbb{Q}$ given by sending $\omega^3 \in \mathfrak{M}_{S^3}$ to $\int_{S^3} \omega^3 \in \mathbb{Q}$. As we said in part B of this chapter, the map $\pi_3(x) \to [I^3(\mathfrak{M}_{(X)})]^*$ defined by sending $[f]$ to $[\int \circ \hat{f}] \in I^3(\mathfrak{M}_X)^*$ gives the duality between $\pi_3(X) \otimes \mathbb{Q}$ and $I^3(\mathfrak{M}_X)$.

Suppose $\omega_1, \ldots, \omega_k \in H^2(X)$ is a basis and $\tilde{\omega}_i$ is a closed 2-form in $A^*(X)$ representing ω_i. Suppose $(\Sigma a_{ij}\, \omega_i \wedge \omega_j,\ \eta) \in P_2[H^2] \oplus A^3(X)$ is a relative cocycle (i.e., $d\eta = \Sigma\, a_{ij}\tilde{\omega}_i \wedge \tilde{\omega}_i$) and that $f: S^3 \to X$. We give a formula for the value of $(\sum_{ij} a_{ij}\omega_i \wedge \omega_j, \eta) \in H^4(P_2[H^2], A^*(X)) = I^3(\mathfrak{M}_X)$ on $[f] \in \pi_3(X)$. To do this, we must deform

$$\mathfrak{M}_X \xrightarrow{\rho_X} A^*(X) \xrightarrow{f^*} A^*(S^3)$$

by homotopy to factor through $\mathfrak{M}_{S^3} = \Lambda(e)$. Define $H\colon P[H^2] \to A^*(S^3) \otimes (t, dt)$ as follows: For each $\omega_i \in H^2$, $f^*\tilde{\omega}_i = d\mu_i$ for some 1-form $\mu_i \in A^*(S^3)$. Define

$$H(\omega_i) = f^*\tilde{\omega}_i \otimes 1 - d(\mu_i \otimes t).$$

Let $\hat{f}\colon \mathfrak{M}_X(2) \to \Lambda(e)$ be trivial, except in degree zero. At this point, we have

$$\begin{array}{ccc} A^*(X) & \xrightarrow{f^*} & A^*(S^3) \\ {\scriptstyle\rho_X}\uparrow & & \uparrow{\scriptstyle\rho_{S^3}} \\ \mathfrak{M}_X(2) & \xrightarrow[\hat{f}]{} & \Lambda(e) \end{array}$$

a homotopy commutative diagram.

We wish to extend this to a homotopy commutative diagram on $\mathfrak{M}_X(3)$. The map $\hat{f}\colon \mathfrak{M}_X(3) \to \Lambda(e)$ must send the class $\sum_{ij} a_{ij}\omega_i \Lambda\ \omega_j, \eta)$ to some multiple $-\cdot e$. (We assume here that e is chosen so that $\int_{S^3} \rho_{S^3}(e)) = 1$.) If the given homotopy is to extend over $\mathfrak{M}_X(3)$, then – must be equal to

$$\begin{aligned} \lambda = & \int_{S^3} f^*\eta - \int_0^1 \int_{S^3} H(\Sigma a_{ij}\omega_i \wedge \omega_j) \\ = & \int_{s^3} f^*\eta - \Sigma a_{ij} \int_0^1 \int_{S^3} \big(f^*\tilde{\omega}_i \otimes 1 - d(\mu_i \otimes t)\big) \wedge \big(f^*\tilde{\omega}_j \otimes 1 - d(\mu_j \otimes t)\big) \\ = & \int_{S^3} f^*\eta \\ & -\sum a_{ij} \int_0^1 \int_{S^3} (f^*\tilde{\omega}_i 1 - d\mu_i \otimes t + \mu_i \otimes dt) \wedge (f^*\tilde{\omega}_j\, 1 - d_{\mu_j} \otimes t + \mu_j \otimes dt). \end{aligned}$$

The only terms in the second part which will contribute to the integral are

$$\begin{aligned} & f^*\tilde{\omega}_i \wedge \mu_j \otimes dt - d\mu_i \wedge \mu_j \otimes tdt + \mu_i \wedge f^*\tilde{\omega}_j \otimes dt - \mu_i \wedge d\mu_j \otimes tdt \\ = & f^*\tilde{\omega}_i \wedge \mu_j \otimes dt - d\mu_i \wedge \mu_j \otimes tdt + \mu_i \wedge f^*\tilde{\omega}_j dt. \end{aligned}$$

Thus,

$$\lambda = \int_{S^3} f^*\eta - \sum a_{ij} \int_{S^3} f^*\tilde{\omega}_i \wedge \mu_j + \mu_i \wedge f^*\tilde{\omega}_j$$

$$\sum \left[a_{ij}(\frac{1}{2} \int_{S^3} d\mu_i \wedge \mu_j) + \mu_i \wedge d_{\mu_j} \right]$$

$$= \int_{S^3} f^*\eta - \sum \left[a_{ij}(\int_{S^3} f^*\tilde{\omega}_i \wedge \mu_j) - \frac{1}{2} d\mu_i \wedge \mu_j + \mu_i \wedge f^*\tilde{\omega}_j \right.$$

$$\left. -\frac{1}{2}\mu_i \wedge d\mu_j \right]$$

$$= \int_{S^3} f^*\eta - \frac{1}{2}\Sigma \left[a_{ij}(\int_{S^3} f^*\tilde{\omega}_i \wedge \mu_j) + \mu_i \wedge f^*\tilde{\omega}_j \right]$$

This formula is a generalization of Whitehead's formula for the Hopf invariant. Whitehead showed that if $f\colon S^3 \to S^2$ is a C^∞ map, then the Hopf invariant H(f) is given by

$$\int_{S^3} f^*\omega \wedge \mu$$

where $\int_{S^2} \omega = 1$ and $d\mu = f^*\omega$. Here, $\pi_3(S^2) \cong \mathbb{Z}$ and the Hopf invariant of an element is its class under this isomorphism.

14.6 Homotopy Theory of Certain 4-Dimensional Complexes

Let X be a space which is homotopy equivalent to $(\vee_{i=1}^T S^2) \cup_\varphi e^4$, where $\varphi\colon S^3 \to \vee_{i=1}^T S^2$. Any simply connected 4-manifold is so represented. The homotopy type of such a space is determined by $[\varphi] \in \pi_3(\vee_{i=1}^T S^2)$. This group is a free abelian group with generators the Whitehead products $[x_i, x_j]$, $i \leq j$ where x_i is the identity map of S^2 onto the ith factor in the wedge. There is another description of the element $[\varphi]$ in terms of the cohomology ring of X. This ring is determined by the symmetric pairing:

$$H^2(X) \otimes H^2(X) \longrightarrow H^4(X) \xrightarrow{\cong} \mathbb{Z},$$

where the last map is evaluation on the fundamental cycle. Since $H^2(X)$ has a natural basis dual to the spheres, the cup product mapping is given by a symmetric integral matrix (λ_{ij}) It turns out that $[\varphi] = \Sigma_{i\leq j}\, \lambda_{ij}[x_i, x_j]$.

This shows that such homotopy types are classified by equivalence classes of symmetric bilinear pairings. The condition that X satisfy Poincaré duality is that the pairing be nonsingular over $\mathbb{Z}$.

It is unknown which pairings arise from closed simply connected smooth 4-manifolds but for closed, simply connected topological 4-manifolds all pairings satisfying Poincaré duality over the integers occur; see [5]. The diffeomorphism classification of simply connected 4-manifolds is much more complicated than the classification up to homotopy equivalence, see [4].

Classifying such complexes up to rational homotopy equivalence is the same as classifying the symmetric pairings up to rational equivalence (including a change of scale in the value group). Thus, if X and Y are two simplicial complexes of this form, then $\mathfrak{M}_X$ and $\mathfrak{M}_Y$ are isomorphic if and only if the pairings are rationally equivalent.

The algebras of piecewise C^∞ forms on X and Y have isomorphic minimal models if and only if the pairings for X and Y are equivalent over the reals (again including an automorphism of the value group). Two symmetric forms are equivalent over the reals if and only if the pairings are of the same rank, have the same dimensional null space, and have the same signature up to sign.

14.7 $\mathbb{Q}$-Homotopy Type of BU_n and U_n

The Grassmannian of n-dimensional complex linear subspaces in $\mathbb{C}^\infty$ is the classifying space for complex n-plane bundles and is denoted BU_n. (Thus, $BU_n = \lim_{N\to\infty} G(n,N)$). Recall that

$$H^*(BU_n; \mathbb{Z}) \cong \mathbb{Z}[c_1, c_2, \ldots, c_n]$$

where $c_\ell \in H^{2\ell}(BU_\eta; Z)$ is the ℓth *Chern class* of the universal vector bundle. This gives a map

$$BU_n \xrightarrow{f} \Pi_{\ell=0}^{n} K(\mathbb{Z}, 2\ell)$$

which induces an isomorphism on $\mathbb{Q}$-cohomology. Since $(K(\mathbb{Z}, 2\ell))_{(0)} = K(\mathbb{Q}, 2\ell)$, it follows that

$$(BU_n)_{(0)} \xrightarrow{f_{(0)}} \prod_{\ell=1}^{n} K(\mathbb{Q}, 2\ell)$$

is a homotopy equivalence.

Remark. This gives a proof of the following rational isomorphisms:

$$\pi_{2i}(BU_n) \otimes \mathbb{Q} \cong \mathbb{Q} \quad \text{for } 2i \leq 2n$$
$$\pi_{2i+1}(BU_n) \otimes \mathbb{Q} = 0. \quad \text{for } 2i + 1 \leq 2n$$

This is the unstable version of the Bott periodicity theorem over $\mathbb{Q}$. In fact, if

$$BU = \lim_{n \to \infty} Gr(n, 2n)$$

then $H^*(BU, \mathbb{Z}) \cong \mathbb{Z}[c_1, c_2, c_3, \ldots,]$. The same argument as above gives that

$$BU_{(0)} \approx \Pi_{\ell=1}^{\infty} K(\mathbb{Q}, 2\ell).$$

Thus,

$$\pi_{2i}(BU) \otimes \mathbb{Q} \approx \mathbb{Q}$$

$$\pi_{2i+1}(BU) \otimes \mathbb{Q} = 0$$

which is the Bott periodicity over $\mathbb{Q}$. Bott periodicity is the result

$$\pi_{2i}(BU) \approx \mathbb{Z}$$
$$\pi_{2i+1}(BU) = 0$$

together with the fact that

$$\frac{c_n}{(n-1)!} : \pi_{2n}(BU) \to \mathbb{Z}$$

is an isomorphism. This divisibility property seems to be closely related to deducing Bott periodicity over the integers from Bott periodicity over the rationals.

Let U_n be the unitary group. Recall that

$$H^*(U_n, \mathbb{Z}) \cong \mathbb{Z}\{x_1, x_3, \ldots, x_{2n-1}\}$$

is an exterior algebra with odd-dimensional generators.

From this, we deduce that the $\mathbb{Q}$-homotopy type:

$$(U_n)_{(0)} \approx \Pi_{\ell=1}^{n} S_{(0)}^{2\ell-1}.$$

Thus, over $\mathbb{Q}$, U_n looks like a product of odd spheres. Moreover, we see that

$$\pi_{2i-1}(U_n) \otimes \mathbb{Q} \cong \mathbb{Q}, \quad \text{for } 1 \leq i \leq n$$

$$\pi_{2i}(U_n) \otimes \mathbb{Q} = 0$$

The stable $\mathbb{Z}$ version for $U = \lim U_{n\to\infty}$ is

$$\pi_{2i-1}(U) \cong \mathbb{Z}$$

$$\pi_{2i}(U) = 0.$$

This is equivalent to the Bott periodicity above using the exact homotopy sequence of the fibration

$$\begin{array}{ccc} U_n & \longrightarrow & S(n, 2n) \\ & & \downarrow \\ & & Gr(n, 2n) \end{array}$$

for n large compared to the dimension, 2i or 2i − 1, in which we are computing the homotopy groups . Here, S(n, 2n) is the Stiefel manifold of n-frames in 2n-space and Gr(n, 2n) is the Grassmannian of n-planes in 2n-space — cf, the exercises

14.8 Products

Let $\mathcal{A}^*$ and $\mathcal{B}^*$ be DGAs. The tensor product $\mathcal{A}^* \otimes \mathcal{B}^*$ naturally receives the structure of a DGA. If $\mathfrak{M}_\mathcal{A}$ and $\mathfrak{M}_\mathcal{B}$ are minimal models for $\mathcal{A}^*$ and $\mathcal{B}^*$, then $\mathfrak{M}_\mathcal{A} \otimes \mathfrak{M}_\mathcal{B}$ is a minimal model for $\mathcal{A}^* \otimes \mathcal{B}^*$. If X and Y are C^∞ manifolds, then there is a map

$$A^*_{C^\infty}(X) \otimes A^*(Y) \longrightarrow A^*_{C^\infty}(X \times Y)$$

which, by the Künneth theorem, induces an isomorphism on cohomology. If X and Y are simplicial complexes and $(X \times Y)'$ is a triangulation of the product cell complex, then there is a map

$$A^*(X) \otimes A^*(Y) \to A^*((X \times Y)')$$

which (again by the Künneth theorem) induces an isomorphism on cohomology.

It follows that $\mathfrak{M}_{X\times Y} \cong \mathfrak{M}_X \otimes \mathfrak{M}_Y$. This should be viewed as a generalization of the Künneth theorem. It includes the Künneth theorem (by taking cohomology). It also includes the rational or real form of the result that $\pi_i(X\times Y) \cong \pi_i(X)\oplus\pi_i(Y)$, since $I(\mathfrak{M}_X \otimes \mathfrak{M}_Y) \cong I(\mathfrak{M}_X) \oplus I(\mathfrak{M}_Y)$.

14.9 Massey Products

Given a space X and classes

$$\alpha \in H^p(X), \beta \in H^q(X), \gamma \in H^r(X)$$

that satisfy

$$\alpha \cup \beta = 0 \text{ in } H^{p+q}(X), \quad \beta \cup \gamma = 0 \text{ in } H^{q+r}(X),$$

there is defined the *Massey triple product*

$$< \alpha, \beta, \gamma > \in H^{p+q+r-1}(X)/(\alpha \cdot H^{q+r-1}(X) + \gamma \cdot H^{p+q-1}(X)).$$

Using differential forms a, b, c to represent the classes α, β, γ, respectively, we write

$$a \wedge b = dg, \quad b \wedge c = dh$$

where g, h are forms of degrees $p + q - 1$, $q + r - 1$. The $p + q + r - 1$ form

$$k = g \wedge c + (-1)^{p-1} a \wedge h$$

is closed and its cohomology class is well-defined modulo

$$\alpha \cdot H^{q+r-1}(X) + \gamma \cdot H^{p+q-1}(X),$$

as may be directly verified by making different choices in the above recipe. The Massey triple product is then the class in the quotient $H^{p+q+r-1}(X)/(\alpha \cdot H^{q+r-1}(X) + \gamma \cdot H^{p+q-1}(X))$.

Now, we consider the special case of compact Kähler manifolds. We recall the following:

Definition. A *Hodge structure of weight* m is given by $\mathbb{Q}$-vector space $H_{\mathbb{Q}}$ together with a *Hodge decomposition*

$$H_{\mathbb{C}} = \oplus_{p+q=m} H^{p,q}$$

$$H^{p,q} = \overline{H}^{q,p}$$

of its complexification $H_{\mathbb{C}} := H_{\mathbb{Q}} \otimes \mathbb{C} = H_{\mathbb{Q}} \otimes_{\mathbb{Q}} \mathbb{C}$.

Suppose now that M is a compact Kãhler manifold (cf. Sect. 7 in Chap. 0 of [8] for the relevant definitions and notations). If we set $A^{p,q}(M) =$ vector space of complex-valued $C^{\infty}(p, q)$ forms on M, and

$$H^{p,q}(M) = \{\varphi \in A^{p,q}(M) : d\varphi = 0\}/dA^{p+q-1}(M) \cap A^{p,q}(M),$$

then it is a basic fact that the subspaces

$$H^{p,q}(M) \subset H^m(M), \quad p+q=m$$

define a Hodge decomposition on $H^m(M) = H^m(M,\mathbb{Q}) \otimes \mathbb{C}$. Consequently, the cohomology $H^m(M)$ of a compact Kähler manifold has a functorial Hodge structure of weight m (functorial means the obvious thing, to be explained momentarily).

Actually, this statement is a consequence of the following lemma ([8], page 149) about forms on compact Kähler manifolds:

Lemma 14.1 (The principle of two types). *On a compact Kähler manifold* M, *suppose that* $\varphi \in A^{p,q}(M)$ *is a form such that*

$$\varphi = d\eta, \quad \eta \in A^{p+q-1}(M).$$

Then, there are forms η' *and* η'' *such that*

$$\varphi = d\eta' \text{ and } \eta' \in A^{p-1,q}(M)$$

and

$$\varphi = d\eta'' \text{ and } \eta'' \in A^{p,q-1}(M).$$

We shall show that the lemma implies that all Massey triple products are zero on a compact Kähler manifold.

For this, we remark first that the cup product

$$H^m(M) \otimes H^n(M) \to H^{m+n}(M)$$

is compatible with the Hodge structures, meaning that

$$H^{p,q}(M) \otimes H^{r,s}(M) \to H^{p+r,q+s}(M).$$

Thus, in forming the Massey Product $<\alpha,\beta,\gamma>$, it suffices to assume that α,β, and γ are themselves homogeneous with respect to the Hodge decomposition; say, $\alpha \in H^{t,s}(X), \beta \in H^{i,j}(X)$, and $\gamma \in H^{u,v}(X)$. We shall construct two closed differential forms φ_1 and φ_2 which are cohomologous and which represent $<\alpha,\beta,\gamma>$ so that $\varphi_1 \in A^{t+i+u,\, s+j+v-1}(X)$ and $\varphi_2 \in A^{t+i+u-1,\, s+j+v}(X)$. In light of the fact that the Hodge decomposition is a direct sum decomposition, this will prove that φ_1 and φ_2 are exact and hence that $<\alpha, \beta, \gamma>= 0$.

Choose closed forms $a \in A^{t,s}(X)$, $b \in A^{i,j}(X)$, and $c \in A^{u,v}(X)$ representing α,β, and γ. We know that $a \wedge b$ is contained in $A^{t+i,j+s}(X)$ and that it is exact. Choose $y \in A^{t+i,j+s-1}(X)$ and $y' \in A^{t+i-1,j+s}(X)$ so that $dy = dy' = a \wedge b$. The form $y - y'$ is closed. Using the Hodge decomposition in degree $t+s+i+j-1$, we can vary y and y' by closed forms preserving the bi-graded type so that $y - y'$

becomes exact. Similarly, we choose $z \in A^{i+u,j+v-1}(X)$ and $z' \in A^{i+u-1,j+v}(X)$ so that $dz = dz' = b \wedge c$ and $z - z'$ is exact. The forms $y \wedge c + (-1)^{t+s-1} a \wedge z$ and $y' \wedge c + (-1)^{t+s-1} a \wedge z'$ both represent the Massey product $< \alpha, \beta, \gamma >$. The first lies in $A^{s+i+u,\, t+j+v-1}(X)$ and the second lies in $A^{s+i+u-1,t+j+v}(X)$. Furthermore, they are cohomologous. This completes the proof that the Massey products vanish on a compact Kähler manifold.

Remark. The above is a special case of the theorem proved in [3] that the rational homotopy type of a compact Kähler manifold is a formal consequence of the cohomology ring (cf. Sect. 14.4 above for the definition of formal). This result, in turn, is generalized to noncompact algebraic varieties in [18].

Chapter 15
Functorality

In Chap. 12, we made explicit the duality between minimal DGAs over the rationals and rational Postnikov towers. We showed that the minimal model of the p.l. forms of a simplicial complex, $A^*(X)$, is dual to the rational Postnikov tower of X. In this chapter, we discuss the functorial properties of this duality.

15.1 The Functorial Correspondence

Let B and C be simply connected simplicial complexes and let $f: B \to C$ be a continuous mapping. For each vertex v of the triangulation of C, let $\mathring{s}t(v)$ be the open star of v. This is the union of all open simplices of C which have v as a vertex. This gives an open cover of C by $\{\mathring{s}t(v)\}_{v \in \text{vertices}(C)}$. If B' is an subdivision of B so that each simplex of B' lies in $f^{-1}(\mathring{s}t(v))$ for some vertex $v \in C$, then there is a simplicial map $\varphi: B' \to C$ which is homotopic to f. Since the points with rational barycentric coordinates are dense in a simplex, we can always choose B' to be a rational subdivision. The inclusion $I: B' \to B$ induces a restriction map on p.l. forms $A^*(B) \overset{I^*}{\hookrightarrow} A^*(B')$.

Suppose that $\rho_B: \mathfrak{M}_B \to A^*(B)$ and $\rho_C: \mathfrak{M}_C \to A^*(C)$ are given minimal models. The composition $I^* \circ \rho_B: \mathfrak{M}_B \to A^*(B')$ is also a minimal model by Theorem 11.5. There is a map $\widehat{f'}: \mathfrak{M}_C \to \mathfrak{M}_B$, unique up to homotopy, so that

$$\begin{array}{ccc} A^*(C) & \xrightarrow{(f')^*} & A^*(B') \\ \uparrow{\scriptstyle \rho_C} & & \uparrow{\scriptstyle I^* \circ \rho_B} \\ \mathfrak{M}_C & \xrightarrow{\widehat{f'}} & \mathfrak{M}_B \end{array}$$

commutes up to homotopy.

P. Griffiths and J. Morgan, *Rational Homotopy Theory and Differential Forms*,
Progress in Mathematics 16, DOI 10.1007/978-1-4614-8468-4_15,

© Springer Science+Business Media New York 2013

Theorem 15.1. *The association* $f \to \hat{f}'$ *defines a function* $[B, C] \to [\mathfrak{M}_C, \mathfrak{M}_B]$ *that preserves compositions and identities.*

Proof. Let us first show that the homotopy class of $\hat{f}: \mathfrak{M}_C \to \mathfrak{M}_B$ depends only on the homotopy class of $f: B \to C$. Suppose we have two simplicial approximations $f': B' \to C$ and $f'': B'' \to C$ where both B' and B'' are rational subdivisions of B. Form the linear cell complex $B \times I$. There is a rational subdivision of the product cell structure on $B \times I$ which agrees with B' at one end and B'' at the other. If we choose this subdivision fine enough, then there is a simplicial map $H: B \times I \to C$ that extends $f' \coprod f''$ on the ends. This leads to a commutative diagram:

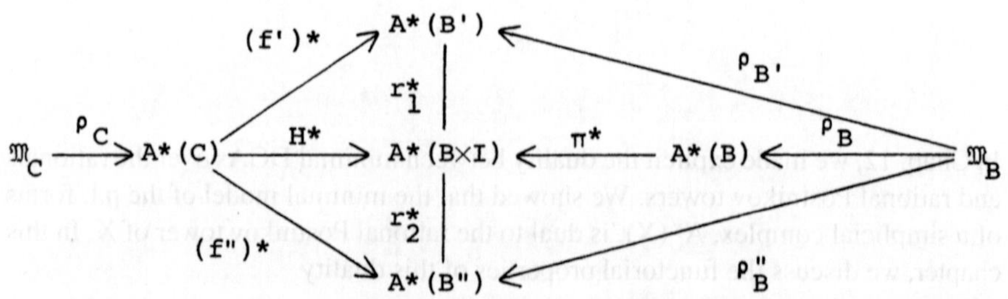

There is $\varphi: \mathfrak{M}_C \to \mathfrak{M}_B$ such that $\pi^* \circ \rho_B \circ \varphi$ is homotopic to $H^* \circ \rho_C$. Thus, $\rho_B \circ \varphi$ is homotopic to $(f')^* \circ \rho_C$, and $\rho_{B''} \circ \varphi$ is homotopic to $(f'')^* \circ \rho_C$. Thus, by 10.8, φ is homotopic to $\hat{f}'$ and $\hat{f}''$. This proves that $[B, C] \to [\mathfrak{M}_C, \mathfrak{M}_B]$ is well defined.

Clearly, $[B, B] \to [\mathfrak{M}_B, \mathfrak{M}_B]$ sends the class of the identity to the class of the identity.

Lastly, suppose

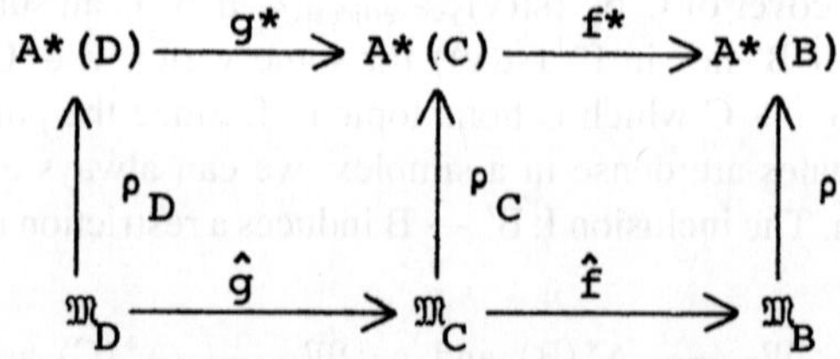

is a diagram in which each square homotopy commutes. Since homotopy is an equivalence relation of the maps from a minimal DGA,

$$\rho_B \circ \hat{f} \circ \hat{g} \cong f^* \circ \rho_C \circ \hat{g} \cong f^* \circ g^{*} \circ \rho_D.$$

Thus, $\hat{f} \circ \hat{g}$ is the map associated to $f^* \circ g^*$. This proves that compositions are preserved. ■

Lemma 15.2. *If* $f: \mathfrak{M} \to \mathfrak{N}$ *is a map between simply connected, minimal DGAs, then it induces a map on indecomposables* $I(f): I(\mathfrak{M}) \to I(\mathfrak{N})$. *As we vary* f *by* *homotopy, the induced map on indecomposables remains unchanged. Thus, there is a well-defined map* $[\mathfrak{M}, \mathfrak{N}] \to \mathrm{Hom}(I(\mathfrak{M}), I(\mathfrak{N}))$.

Proof. Let $H\colon \mathfrak{M} \longrightarrow \mathfrak{N} \otimes (t, dt)$ be a homotopy. Since $\mathfrak{M}$ and $\mathfrak{N}$ are minimal and have no generators in degree 1, it follows that H induces a map:

$$H\colon \mathfrak{M} \wedge \mathfrak{M} \to (\mathfrak{N} \wedge \mathfrak{N}) \otimes (t, dt).$$

For any $v \in \mathfrak{M}$, we have $dv \in \mathfrak{M} \wedge \mathfrak{M}$. We write

$$H(v) = v_0 + \sum_{i \geq 1} v_i \otimes t^i + w_i \otimes t^{i-1}dt.$$

Then,

$$dH(v) = H(dv) = dv_0 + \sum_{i \geq 1} dv_i \otimes t^i + (\pm i v_i + dw_i) \otimes t^{i-1}dt.$$

This expression and $dw_i \otimes t^{i-1}dt$ are each contained in $(\mathfrak{N} \wedge \mathfrak{N}) \otimes (t, dt)$. Thus, so is $\sum_{i \geq 1} iv_i \otimes t^{i-1}dt$. It follows that for every $i \geq 1$, we have $v_i \in \mathfrak{N} \wedge \mathfrak{N}$. Consequently, the maps induced by H at $t = 0$ and $t = 1$ from $I(\mathfrak{M})$ to $I(\mathfrak{N})$ are the same. ∎

As an example of Theorem 15.1, let $B = S^n$. We have a map:

$$[S^n, X] = \pi_n(X) \longrightarrow [\mathfrak{M}_X, \mathfrak{M}_{S^n}] \cong \mathrm{Hom}(I(\mathfrak{M}_X), I(\mathfrak{M}_{S^n})).$$

There is an isomorphism $I^n(\mathfrak{M}_{S^n}) \to \mathbb{Q}$ given by sending

$$\alpha \in \mathfrak{M}_{S^n} \text{ to } \int_{S^n} \alpha \in \mathbb{Q}.$$

Thus, we have a map:

$$\pi_n(X) \otimes \mathbb{Q} \xrightarrow{\lambda_X} \mathrm{Hom}(I^n(\mathfrak{M}_X), \mathbb{Q}) = [I^n(\mathfrak{M}_X)]^*.$$

This map agrees with the identification given in Chap. 12. If $f\colon X \to Y$ is given and $\hat{f}\colon \mathfrak{M}_Y \to \mathfrak{M}_X$ is an associated map on minimal models, then the following diagram commutes:

$$\begin{array}{ccc}
\pi_n(X) \otimes \mathbb{Q} & \xrightarrow[\lambda_X]{} & I^n(\mathfrak{M}_X)^* \\
\Big\downarrow f_* \otimes \mathrm{Id}_{\mathbb{Q}} & & \Big\downarrow I^n(\hat{f}^*) \\
\pi_n(y) \otimes \mathbb{Q} & \xrightarrow[\lambda_Y]{} & I^n(\mathfrak{M}_Y)^*
\end{array}$$

This follows immediately from the fact that the correspondence in 15.1 is natural and preserves compositions. We shall show later in this chapter that λ_X is an isomorphism for all simply connected X.

15.2 Bijectivity of Homotopy Classes of Maps

We begin now the proof that if C is a local space, then the map $[B, C] \longrightarrow [\mathfrak{M}_C, \mathfrak{M}_B]$ given in Theorem 15.1 is a bijection. This is proved by induction on the stages of the Postnikov tower of C. The inductive step requires a detailed analysis of a long exact sequence. It is this analysis that the next 5 propositions develop.

Proposition 15.3. *Let* $p\colon E \to C_0$ *be a simplicial map with homotopy theoretic fiber* $K(\pi, n)$. *Suppose* $\pi_1(C_0) = \{e\}$. *Suppose that* $V = \pi \otimes \mathbb{Q}$ *is a finite-dimensional rational vector space. Let* $\rho_{C_0}\colon \mathfrak{M}_{C_0} \to A^*(C_0)$ *be a minimal model. Let* $\mathfrak{M}' = \mathfrak{M}_{C_0} \otimes_d \Lambda(V^*)^n$ *and suppose there is a map* $\rho\colon \mathfrak{M}' \to A^*(E)$ *such that:*

(1) $\rho|\mathfrak{M}_{C_0} = p^* \circ \rho_{C_0}$.
(2) ρ *induces an isomorphism on rational cohomology.*

Then, there is a commutative diagram:

$$\begin{array}{ccc}
[X, C_0] & \xrightarrow[\textit{lifting}]{\textit{obst. to}} & H^{n+1}(X; \pi) \\
\Big\downarrow & & \Big\downarrow \otimes\mathbb{Q} \\
 & & H^{n+1}(X; V) \\
 & & \Big\downarrow \rho_X^* \\
[\mathfrak{M}_{C_0}, \mathfrak{M}_X] & \xrightarrow[\textit{extending}]{\textit{obst. to}} & H^{n+1}(\mathfrak{M}_X; V).
\end{array}$$

Proof. The obstruction to lifting $f\colon X \to C_0$ is $f^*(k)$ where $k \in H^{n+1}(C_0; \pi)$ is the k-invariant of the fibration. The obstruction to extending $\varphi\colon \mathfrak{M}_{C_0} \to \mathfrak{M}_X$ over $\mathfrak{M}'$ is $\varphi^*([d]) \in H^{n+1}(\mathfrak{M}_X; V)$. As we saw in 11.5, [d] is the class of $k \otimes 1 \in H^{n+1}(C_0; \pi) \otimes \mathbb{Q}$ under the identification $\rho_{C_0}^*\colon H^*(\mathfrak{M}_{C_0}) \longrightarrow H^*(C_0)$. If f and φ correspond, then $f^* = \varphi^*$ under the identifications $\rho_{C_0^*}\colon H^*(\mathfrak{M}_{C_0}) \xrightarrow{\sim} H^*(C_0)$. and $\rho_{C_0^*}\colon H^*(\mathfrak{M}_X) \xrightarrow{\sim} H^*(X)$. The proposition is immediate from these facts. ■

Proposition 15.4. *Let* $p\colon C_{n+1} \to C_n$ *be a principal fibration with fiber* $K(\pi, n+1)$. *For any CW-complex* X, *there is an exact sequence:*

$$H^{n+1}(X; \pi) \longrightarrow [X, C_{n+1}] \xrightarrow{p_*} [X, C_n] \xrightarrow{\theta} H^{n+2}(X; \pi).$$

This means that $\mathrm{Im}(p_*) = \theta^{-1}(0)$, *and that there is an action of* $H^{n+1}(X;\pi)$ *on* $[X, C_{n+1}]$ *so that two maps are in the same orbit if and only if they have the same image in* $[X, C_n]$. *The isotropy group of this action at* $f \in [X, C_{n+1}]$ *is the subgroup* $H_f \subset H^{n+1}(X;\pi)$ *consisting of all elements* α *for which there is a map* $X \times S^1 \xrightarrow{\varphi} C_n$ *with* $\varphi|X \times p = p \circ f$ *and the obstruction to lifting* φ *to* C_{n+1} *is* $\alpha \otimes \iota \in H^{n+2}(X \times S^1;\pi)$ *where* $\iota \in H^1(S^1)$ *is the generator.*

Proof. Since θ is the obstruction to lifting, clearly, $\theta^{-1}(0) = \mathrm{Im}(p_*)$.

Let us define the action of $H^{n+1}(X;\pi)$ on $[X, C_{n+1}]$. Given $\alpha \in H^{n+1}(X;\pi) = H^{n+2}(X \times I, X \times \partial I;\pi)$ and $f: X \to C_{n+1}$, define a map $\alpha \cdot f: X \to C_{n+1}$ such that $p \circ \alpha \cdot f = p \circ f$ and the obstruction to finding a homotopy of liftings of $p \circ f$ connecting f to $\alpha \cdot f$ is α. One checks easily that this is an action of $H^{n+1}(X;\pi)$ on $[X, C_{n+1}]$. Clearly, $p \circ f = p \circ g$ if and only if $[f] = \alpha \cdot [g]$ for some $\alpha \in H^{n+1}(X;\pi)$.

Suppose $\alpha \cdot f = f$ in $[X, C_{n+1}]$. Let $H: X \times I \to C_{n+1}$ be a homotopy form f to αf. Since $p \circ f = p \circ \alpha f$, the composition $p \circ H$ defines a map $\varphi: X \times S^1 \to C_n$. Clearly, $\varphi|X \times \{0\} = p \circ f$. We claim that the obstruction to lifting φ to C_{n+1} is $\alpha \otimes \iota \in H^{n+2}(X \times S^1;\pi)$. Since we have a lifting $H: X \times I \to C_{n+1}$, the obstruction to lifting φ as a map of $X \times S^1$ lies in $H^{n+1}(X;\pi) \otimes H^1(S^1;\mathbb{Z}) \subset H^{n+2}(X \times S^1;\pi)$. It is easily identified with $\alpha \otimes \iota$. ■

There is an analogous sequence from Hirsch extensions of DGAs which we construct now. Let $\mathfrak{M}' = \mathfrak{M} \otimes_d \Lambda(V^*)^{n+1}$ be a Hirsch extension, with V being a finite-dimensional rational vector space. Let $\mathcal{A}$ be a DGA.

Proposition 15.5. *There is an exact sequence:*

$$H^{n+1}(\mathcal{A};V) \longrightarrow [\mathfrak{M}', \mathcal{A}] \xrightarrow{i^*} [\mathfrak{M}, \mathcal{A}] \xrightarrow{\theta} H^{n+2}(\mathcal{A};V).$$

This means that $\theta^{-1}(0) = \mathrm{Im}\, i^*$ *and that the group* $H^{n+1}(\mathcal{A};V)$ *acts on* $[\mathfrak{M}', \mathcal{A}]$ *so that two elements are in the same orbit if and only if they have the same image in* $[\mathfrak{M}, \mathcal{A}]$. *The isotropy group for this action at* $\varphi \in [\mathfrak{M}', \mathcal{A}]$ *is the subgroup of all* $\alpha \in H^{n+1}(\mathcal{A}, V)$ *for which there is a map* $\psi: \mathfrak{M} \to \mathcal{A} \otimes \Lambda(e)$, *with* $p_{\mathcal{A}} \circ \psi = \varphi|\mathfrak{M}$ *(where* $p_{\mathcal{A}}$ *is the projection to* $\mathcal{A}$ *obtained by setting* $e = 0$*) with the obstruction to extending* ψ *over* $\mathfrak{M}'$ *being* $\alpha \otimes e \in H^{n+2}(\mathcal{A} \otimes \Lambda(e);V)$.

Proof. Since θ is the obstruction to extending over $\mathfrak{M}'$, $\theta^{-1}(0) = \mathrm{Im}\, i^*$.

Let us define the action of $H^{n+1}(\mathcal{A};V) = \mathrm{Hom}(V^*, H^{n+1}(\mathcal{A}))$ on $[\mathfrak{M}', \mathcal{A}]$. Given $\alpha: V^* \to H^{n+1}(\mathcal{A})$ and $\varphi: \mathfrak{M}' \to \mathcal{A}$, we define $\alpha \cdot \varphi$. On $\mathfrak{M} \subset \mathfrak{M}'$, $\alpha \cdot \varphi = \varphi$. Choose a lifting $\tilde{\alpha}: V^* \to \{\text{closed forms of } \mathcal{A}^{n+1}\}$ for α. Define $\tilde{\alpha} \cdot \varphi: V^* \to \mathcal{A}$ by $\alpha \cdot \varphi(v) = \varphi(v) + \tilde{\alpha}(v)$. One sees easily that this defines a map of DGAs. The homotopy class of $\tilde{\alpha} \cdot \varphi$ depends only on the homotopy class of φ and the cohomology class of $\tilde{\alpha}$. Let us show that φ_1, φ_2 have the same image under i^* if and only if there is $\alpha \in H^{n+1}(\mathcal{A}, V)$ so that $\alpha\varphi_1 = \varphi_2$. The "if" direction is immediate from the definition of the action. Suppose, conversely, that we have maps φ_1 and φ_2 so that $\varphi_1|\mathfrak{M}$ is homotopic to $\varphi_2|\mathfrak{M}$. Consider the diagram:

$$
\begin{array}{ccc}
\mathfrak{M} & \xrightarrow{\varphi_1|\mathfrak{M}} & \mathcal{A} \\
\downarrow & & \downarrow = \\
\mathfrak{M} \otimes_d \Lambda(V^*) & \xrightarrow[\varphi_2]{} & \mathcal{A}
\end{array}
$$

It is homotopy commutative. Since $H^*(\mathcal{A}, \mathcal{A}) = 0$, there is no obstruction to finding $\varphi_2' : \mathfrak{M}' \to \mathcal{A}$ which is homotopic to φ_2 and such that $\varphi_2'|\mathfrak{M} = \varphi_1|\mathfrak{M}$.

Thus, we may assume that $\varphi_1|\mathfrak{M} = \varphi_2|\mathfrak{M}$. For each $v \in V^*$, consider $\varphi_1(v) - \varphi_2(v) \in \mathcal{A}^{n+1}$. Since $dv \in \mathfrak{M}$, we have $d\varphi_1(v) = \varphi_1(dv) = \varphi_2(dv) = d\varphi_2(v)$. Hence, $\varphi_1(v) - \varphi_2(v)$ is closed. If $\alpha : V^* \to H^{n+1}(\mathcal{A})$ is the resulting homomorphism, then one sees easily that $\alpha \circ \varphi_2 = \varphi_1$ in $[\mathfrak{M}', \mathcal{A}]$.

Lastly, we need to understand the isotopy groups of this action. Let $\tilde{\alpha} : V \to$ {closed forms of $\mathcal{A}^{n+1}$} be a representative for α. Let $\varphi : \mathfrak{M}' \to \alpha$. If $\tilde{\alpha} \circ \varphi = \varphi$ in $[\mathfrak{M}', \mathcal{A}]$, then there is a homotopy $H : \mathfrak{M}' \to \mathcal{A} \otimes (t, dt)$ from φ to $\tilde{\alpha} \circ \varphi$. If we restrict H to $\mathfrak{M}$, it becomes a homotopy from $\varphi|\mathfrak{M}$ to $\varphi|\mathfrak{M}$.

Let $K \subset (t, dt)$ be all forms $\Sigma\, a_i t^i + b_i t^i dt$ such that $\Sigma_{i \geq 1} a_i = 0$. This is a sub-DGA. The algebra $\mathcal{A} \otimes K \subset \mathcal{A} \otimes (t, dt)$ is the kernel of $r_1 - r_0 : \mathcal{A} \otimes (t, dt) \to \mathcal{A}$ where r_i is restriction at $t = i$. ■

Lemma 15.6. $\Lambda(dt) \subset K$ *induces an isomorphism on cohomology.*

This is a straightforward computation.

With this lemma in place, let us return to the proof of the proposition. As a consequence of the lemma, we have that $j : \mathcal{A} \otimes \Lambda(e) \to \mathcal{A} \otimes K$, defined by $e \to dt$, induces an isomorphism on cohomology.

Consider the diagram:

$$
\begin{array}{ccccc}
 & & \mathcal{A} \otimes \Lambda(e) & & \\
 & \overset{H'}{\nearrow} & \downarrow j & \overset{p_{\mathcal{A}}}{\searrow} & \\
\mathfrak{M} & \xrightarrow{H} & \mathcal{A} \otimes K & \xrightarrow{r} & \mathcal{A}
\end{array}
$$

where $r : \mathcal{A} \otimes K \to \mathcal{A}$ is restriction at $t = 0$, and $p_{\mathcal{A}}$ is defined by setting $e = 0$. The obstructions to lifting H up to homotopy vanish. Thus, there is $H' : \mathfrak{M} \to \mathcal{A} \otimes \Lambda(e)$ so that $p_{\mathcal{A}} \circ H' = \varphi$ and so that $j \circ H'$ is homotopic to H.

There is no obstruction to extending H' to a homotopy $J' : \mathfrak{M}' \to \mathcal{A} \otimes (t, dt)$ such that:

(1) $J'|\mathfrak{M} = H'$ and
(2) J' is a homotopy from φ to $\tilde{\alpha} \circ \varphi$.

We claim that the obstruction to extending H' to a map $\mathfrak{M}' \to \mathcal{A} \otimes \Lambda(e)$ is exactly the homomorphism $v \to [\tilde{\alpha}(v) \otimes e] \in H^{n+2}(\mathcal{A} \otimes \Lambda(e))$. To see this, note that the obstruction to extending H' sends $v \in V$ to the class $H'(dv) \in H^{n+2}(\mathcal{A} \otimes \Lambda(e))$.

Since $J' \colon \mathfrak{M}' \to \mathcal{A} \otimes (t, dt)$ is an extension of $j \circ H'|\mathfrak{M}$, we have $H'(dv) = J'(dv) = dJ'(v)$. Thus, $j \circ H'(dv)$ is of the form $a_v \otimes 1 + b_v \otimes dt$. Suppose $J'(v) = \sum a_{v,i} \otimes t^i + b_{v,i} \otimes t^i dt$. We see that

$$\sum da_{v,i} \otimes t^i + (-1)^{\deg(a_{v,i})} i a_{v,i} \otimes t^{i-1} dt + db_{v,i} \otimes t^i dt$$
$$= a_v \otimes 1 + b_v \otimes dt.$$

This implies that $a_v = da_{v,0}$ and hence that a_v is exact. Since

$$d \int_{t=0}^{t=1} H(v) + \int_{t=0}^{t=1} H(dv) = H|_{t=1}(v) - H|_{t=0}(v) = \tilde{\alpha}(v)$$

we have

$$(-1)^{\deg(b_v)} b_v = \tilde{\alpha}(v) - d\left(\int_{t=0}^{t=1} H(v)\right).$$

Thus, in cohomology, the class of $H'(dv)$ is $(-1)^{\deg(b_v)}(\tilde{\alpha}(v) \otimes e)$. This proves that, up to sign, the obstruction to extending H' is the homomorphism $v \to [\tilde{\alpha}(v) \otimes e] \in H^{n+2}(\mathcal{A} \otimes \Lambda(e))$. This completes the proof of Proposition 15.5.

Let $p \colon C_{n+1} \to C_n$ be a simplicial model for a principal fibration with fiber $K(V, n+1)$, where V is a finite-dimensional rational vector space. Let $\varphi \colon \mathfrak{M} \to A^*(C_n)$ be a minimal model. Let $\mathfrak{M}' = \mathfrak{M} \otimes_d \Lambda(V^*)$ and let $\varphi' \colon \mathfrak{M}' \to A^*(C_{n+1})$ be a map extending $p^* \circ \varphi$ and inducing an isomorphism on cohomology. Let X be a simplicial complex, and let $\rho_X \colon \mathfrak{M}_X \to A^*(X)$ be a minimal model. There is a commutative diagram of exact sequences:

$$\begin{array}{ccccccc}
H^{n+1}(X;V) & \longrightarrow & [X, C_{n+1}] & \longrightarrow & [X, C_n] & \longrightarrow & H^{n+2}(X;V) \\
\downarrow{\scriptstyle \rho_X^*} & & \downarrow & & \downarrow & & \downarrow{\scriptstyle \rho_X^*} \\
H^{n+1}(\mathfrak{M}_X;V) & \longrightarrow & [\mathfrak{M}', \mathfrak{M}_X] & \longrightarrow & [\mathfrak{M}, \mathfrak{M}_X] & \longrightarrow & H^{n+2}(\mathfrak{M}_X;V)
\end{array}$$

We check now that the action of $H^{n+1}(X;V)$ on $[X, C_{n+1}]$ and action of $H^{n+1}(\mathfrak{M}_X;V)$ on $[\mathfrak{M}', \mathfrak{M}_X]$ are compatible.

Suppose $\alpha \in H^{n+1}(X, V)$ and $f \colon X \to C_{n+1}$ are given. The maps

$$\mathfrak{M}' \xrightarrow{\varphi'} A^*(C_{n+1}) \xrightarrow{f^*} A^*(X)$$

and

$$\mathfrak{M}' \xrightarrow{\varphi'} A^*(C_{n+1}) \xrightarrow{(\alpha f)^*} A^*(X)$$

are homotopic on $\mathfrak{M}$. The obstruction to extending the homotopy over $\mathfrak{M}'$ is exactly $\alpha \in H^{n+1}(X; V)$. Thus, if φ_f and $\varphi_{\alpha f}$ are maps $\mathfrak{M}' \to \mathfrak{M}_X$ representing f and αf, then they are homotopic on $\mathfrak{M}$, and the obstruction to extending the homotopy over $\mathfrak{M}'$ is α. Hence, $\varphi_{\alpha f} = \alpha \circ \varphi_f$ in $[\mathfrak{M}', \mathfrak{M}_X]$. This completes the proof of Proposition 15.5

Theorem 15.7. *Let* C *be a local space with homotopy groups that are finite-dimensional rational vector spaces, and let* X *be a simplicial complex. The functor* $[X, C] \to [\mathfrak{M}_C, \mathfrak{M}_X]$ *is a bijection.*

Proof. The proof is by induction on the Postnikov tower of C. We show that $[X, C_n] \to [\mathfrak{M}_C(n), \mathfrak{M}_X]$ is a bijection for all n, where C_n is the nth stage in the Postnikov tower for C and $\mathfrak{M}_C(n)$ is the subalgebra of $\mathfrak{M}_C$ generated in degrees $\leq$ n. The inductive step follows by the 5 lemma from the commutative diagram above, together with the fact that the identification $H^{n+1}(X, V) \xrightarrow{\rho_X^*} H^{n+1}(\mathfrak{M}_X; V)$ induces group isomorphisms on the corresponding isotropy groups. ∎

Corollary 15.8. *The functorial mapping* $\lambda_X : \pi_n(X) \otimes \mathbb{Q} \to I^n(\mathfrak{M}_X)^*$ *is an isomorphism for all simply connected spaces* X *with finite-dimensional rational cohomology in every degree.*

Proof. We show that λ_X is onto. Let $\varphi: I^n(\mathfrak{M}_X) \to \mathbb{Q}$ be given. There is a map of DGA s $\hat{\varphi}: \mathfrak{M}_X \to \mathfrak{M}_{S^n}$ which realizes $\hat{\varphi}$. It is clear how to define $\hat{\varphi}$ in degrees $\leq$ n. In the higher degrees, there are no obstructions to extending $\hat{\varphi}$ since $H^*(\mathfrak{M}_{S^n}) = 0$ for $* >$ n. Once we have $\hat{\varphi}$, Theorem 15.7 tells us there is $\varphi: S^n \to X_{(0)}$ which realizes $\hat{\varphi}$. Thus, $\lambda_{X_{(0)}} : \pi_n(X_{(0)}) \longrightarrow I^n(\mathfrak{M}_X^*)$ is onto. But $\pi_n(X_{(0)}) = \pi_n(X) \otimes \mathbb{Q}$, and by naturality $\lambda_{X_{(0)}} = \lambda_X \otimes Id_{\mathbb{Q}}$.

To see that λ_X is 1-1, suppose $\lambda_X(f) = 0$. This implies that $\hat{f} : \mathfrak{M}_X \to \mathfrak{M}_{S^n}$ induces the zero map on indecomposables in degree n. Elementary obstruction theory implies that $\hat{f}$ itself is homotopic to zero. Hence, by 15.7, the composition $S^n \xrightarrow{f} X \subset X_{(0)}$ is homotopically trivial. This means that $[f] \in \pi_n(X)$ is trivial in $\pi_n(X_{(0)}) = \pi_n(X) \otimes \mathbb{Q}$. ∎

Corollary 15.9. *Let* X *and* Y *be simply connected spaces, each with finite-dimensional rational cohomology in every degree, and let* $f: X \to Y$ *a continuous map. From* $\mathfrak{M}_X$ *and* $\mathfrak{M}_Y$, *we can construct spaces* $X_{(0)}$ *and* $Y_{(0)}$ *which are localizations for* X *and* Y. *From* $\hat{f} \in [\mathfrak{M}_Y, \mathfrak{M}_X]$, *we can construct a map* $f_{(0)}: X_{(0)} \to Y_{(0)}$ *which is a representative, up to homotopy, of the localization of* f.

15.3 Equivalence of Categories

All of the results of this section can be formalized by saying that a functor between two categories is an equivalence of categories.

Definition. Let $\mathcal{C}_1$ and $\mathcal{C}_2$ be categories and let $F: \mathcal{C}_1 \to \mathcal{C}_2$ be a functor. F is said to be an *equivalence of categories* if the following hold:

1. Every object B of $\mathcal{C}_2$ is isomorphic to an object F(A) for some object A of $\mathcal{C}_1$.
2. For all objects A_1 and A_2 of $\mathcal{C}_1$ the map induced by F

$$\mathrm{Hom}_{\mathcal{C}_1}(A_1, A_2) \longrightarrow \mathrm{Hom}_{\mathcal{C}_2}(F(A_1), F(A_2))$$

is a bijection.

Given a category $\mathcal{C}$ and a set of morphisms S of $\mathcal{C}$ that include all isomorphisms and are closed under compositions, we define the localization of $\mathcal{C}$ at S by formally inverting all morphisms in S. Thus, the objects of the new category are the same as the objects of $\mathcal{C}$, but the morphisms from A to B in the new category consist of a string of morphisms:

$$A = A_0 \longrightarrow A_1 \longleftarrow A_2 \cdots \longrightarrow A_k = B,$$

where all the arrows to the 'left' are elements of S. If successive arrows in the same direction can be replaced by the composite arrow, then by definition the morphism is unchanged.

The first category with which we shall deal is the rational homotopy category of simply connected spaces of finite type. We begin with the category whose objects are simply connected, simplicial complexes whose rational homology groups are finite dimensional in each degree and whose morphisms are homotopy classes of maps. Then, we localize this category by localizing at those maps that induce isomorphisms on rational homology or equivalently on rational homotopy. We denote this category by $\mathbf{HOM}_{(0)}$. The other category is the category of simply connected DGAs over $\mathbb{Q}$ with finite-dimensional cohomology in each degree, localized at the collection of DGA maps that induce isomorphisms on cohomology. This category is denoted $\mathbf{DGA}_{(0)}$ We have a functor "p.l. forms" from $\mathbf{HOM}_{(0)}$ to $\mathbf{DGA}_{(0)}$. The existence of a minimal model shows that every object of $\mathbf{DGA}_{(0)}$ is isomorphic to a minimal DGA. Associated to a minimal DGA, there is a rational Postnikov tower or equivalently a simplicial complex whose p.l. forms have a minimal model isomorphic to the given one. This shows that every object of $\mathbf{DGA}_{(0)}$ is isomorphic to the DGA of p.l. forms on some simplicial complex. The fact that the p.l. forms functor induces a bijection on the sets of morphisms follows easily from Theorem 15.7. This shows that the functor of p.l. forms induces an isomorphism from $\mathbf{HOM}_{(0)}$ to $\mathbf{DGA}_{(0)}$. That is to say, the rational homotopy category of simply connected spaces of finite type is isomorphic to the homotopy category of simply connected DGAs over $\mathbb{Q}$ with finite-dimensional cohomology in each degree, and the isomorphism between these rational homotopy categories is given by associating to a simplicial complex is p.l. forms.

Chapter 16
The Hirsch Lemma

The purpose of this chapter is to prove Theorem 12.1. Namely, we shall show the following. Let $E \to B$ be a principal $K(\pi, n)$-fibration, and let $f': E' \to B'$ be a simplicial model for it. Given a minimal model $\mathfrak{M}$ for the p.l. forms on B', the Hirsch extension of $\mathfrak{M}$ dual to the fibration is a minimal for the p.l. forms on E'.

16.1 The Cubical Complex and Cubical Forms

Definition. Let X be a topological space. For each $q \geq 0$, let I^q be the unit cube in $\mathbb{R}^q$, namely, $I^q = \prod_{i=1}^{q}[0,1]$. For any $0 \leq i \leq q$ and $\epsilon \in \{0,1\}$, we define the (i,ϵ)-*face* of I^q denoted $f_i^\epsilon(I^q)$ the subset where i^{th} coordinate is equal to ϵ. This face is identified with I^{q-1} by the map

$$f_i^\epsilon : (x_1, \ldots, x_{q-1}) \mapsto (x_1, \ldots, x_{i-1}, \epsilon, x_i, \ldots, x_{q-1}).$$

We define the *boundary* of I^q by

$$\partial I^q = \sum_{\epsilon=0}^{1} \sum_{i=1}^{q} (-1)^i (-1)^\epsilon f_i^\epsilon,$$

considered as a formal linear combination of maps of the I^{q-1}-cube into I^q.

A *singular* q-*cube* in X is a continuous map $I^q \to X$. A *singular cube* in X is a singular q-cube in X for some $0 \leq q < \infty$. For any $q \geq 0$, we denote the set of singular q-cubes in X by $C_q(X)$. Let σ be a singular q-cube in X. For any $0 \leq i \leq q$ and $\epsilon \in \{0,1\}$, we define the (i,ϵ)-*face* of σ, denoted $f_i^\epsilon(\sigma)$ to be the composition

$$I^{q-1} \xrightarrow{f_i^\epsilon} I^q \xrightarrow{\sigma} X.$$

Lastly, we say that a q-cube σ in X is *degenerate* if σ factors through the map $I^q \xrightarrow{p_q} I^{q-1}$ defined by $p_q(x_1, \ldots, x_q) = (x_1, \ldots, x_{q-1})$. (Notice that degeneracies

P. Griffiths and J. Morgan, *Rational Homotopy Theory and Differential Forms*,
Progress in Mathematics 16, DOI 10.1007/978-1-4614-8468-4_16,
© Springer Science+Business Media New York 2013

are defined only with respect to the last coordinate.) We denote by $\hat{Q}(X)$ the chain complex whose q^{th} chain group is the free abelian group generated by the set $C_q(X)$ and with the boundary map given by

$$\partial\sigma = \sum_{\epsilon=0}^{1}\sum_{i=1}^{q}(-1)^i(-1)^\epsilon f_i^\epsilon(\sigma).$$

It is easy to see that the boundary of a degenerate cube is a linear combination of degenerate cubes of one lower dimension, so that the degenerate cubes generate a subcomplex. We denote by $Q(X)$ the quotient chain complex $\hat{Q}(X)$ by the subchain complex generated by the degenerate cubes.

Lemma 16.1. *The homology of the chain complex* $Q(X)$ *is naturally identified with the singular homology of* X.

Proof. Just as in the case of singular homology, one shows that all the axioms for homology are verified. [Excision requires subdivision of a cube into smaller cubes. The computation of the homology of a point is straightforward, but it is worth noting that the homology of a point computed using $\hat{Q}$ is isomorphic to $\mathbb{Z}$ in every degree since the boundary of a cube has an even number of faces, half counted with a positive sign and half counted with a negative sign. Thus, it is crucial that we set degenerate cubes equal to zero in order to compute singular homology.] ■

Given a topological space X, denote by $|\hat{Q}^q(X)|$ the disjoint union over $\sigma \in C_q(X)$ of I^q. A point of $|\hat{Q}^q(X)|$ is denoted $(x_1, \ldots, x_q; \sigma)$, with the x_i being in the interval $[0, 1]$ and σ being an element of $C_q(X)$. We define

$$|\hat{Q}(X)| = \coprod_{q=0}^{\infty} |\hat{Q}^q(X)|.$$

We introduce an equivalence relation on $|\hat{Q}(X)|$ as follows. It is generated by two elementary equivalences:

1. For $\sigma \in C_q(X)$ and any $(x_1 \ldots, x_{q-1}) \in I^{q-1}$, we identify $(x_1, \ldots, x_{q-1}; f_i^\epsilon(\sigma))$ with $(x_1 \ldots, x_{i-1}, \epsilon, x_i, \ldots, x_{q-1}; \sigma)$.
2. For any $\sigma \in C_q(X)$ if $\sigma = \tau \circ p_q$, then for any $(x_1, \ldots, x_q) \in I^q$, we identify $(x_1, \ldots, x_q; \sigma)$ with $(x_1, \ldots, x_{q-1}; \tau)$.

The quotient space is denoted $|Q(X)|$. Notice that this space is a union of compact subsets which are the images of the various closed cubes in $|\hat{Q}|$ and is a CW complex with one cell for each nondegenerate singular cube in X. The closures of the open cubes are not necessarily embeddings of the closed cube because some of the faces of a nondegenerate cube can be degenerate and hence collapsed to lower dimensional cubes in $|Q(X)|$. Still, the topology of $|Q(X)|$ is generated by the images of the closed cubes, meaning that a subset of $|Q(X)|$ is open if and only if its intersection with the image of each closed cube is an open subset of that image. There is the obvious map from $|\hat{Q}(X)| \to X$ which is compatible with the identifications and hence factors to give a continuous map $|Q(X)| \to X$. The fact that the chain complex Q(X) computes the homology of X implies that this map is a weak homotopy equivalence; i.e., this map induces an isomorphism on the homology and homotopy groups. If X is the homotopy type of a CW complex, then it is a homotopy equivalence.

There is a DGA, denoted $\Omega^*(X)$, defined over $\mathbb{Q}$, of piecewise linear forms on $|Q(X)|$. One way to think about this is that associated to the cube I^q, we have the DGA, denoted $\Omega^*(I^q)$, of polynomial forms on the cube with rational coefficients $\Lambda^*(x_1, \ldots, x_q, dx_1, \ldots, dx_q)$. Then an element of $\Omega^n(X)$ is a collection of forms $\{\omega_\sigma\}_\sigma$ indexed by $\sigma \in \coprod_q C_q(X)$ where $\omega_\sigma \in \Omega^n(I^q)$ when σ is a singular q-cube in X. This collection is required to satisfy two compatibility conditions:

1. $\omega_\sigma|_{f_i^\epsilon(\sigma)}$ is identified with $\omega_{f_i^\epsilon(\sigma)}$ under the natural identification of I^{q-1} with $f_i^\epsilon(I^q)$.
2. If $\sigma = \tau \circ p_q$, then $\omega_\sigma = p_q^*(\omega_\tau)$.

In an obvious way, $\Omega^*(X)$ can be viewed as piecewise polynomial forms on the cubical complex $|Q(X)|$.

Since $|Q(X)|$ is a cell complex, it has cellular chains: There is a generator in degree q for each nondegenerate singular q-cube in X with the boundary map induced by

$$\partial I^q = \sum_{\epsilon=0}^{1}\sum_{i=0}^{q}(-1)^i(-1)^\epsilon f_i^\epsilon(I^q).$$

This chain complex computes the homology of $|Q(X)|$ and hence of X. The following result is proved along the same lines as the corresponding result for p.l. forms on a simplicial complex:

Proposition 16.2. *Integration defines a map from $\Omega^*(X)$ to the cellular cochains on $|Q(X)|$ with rational coefficients. This map induces a ring isomorphism from the cohomology of $\Omega^*(X)$ to the singular cohomology of $|Q(X)|$. Hence, the cohomology of $\Omega^*(X)$ is identified with the singular rational cohomology* X.

We can associate a simplicial complex S(X) to $|Q(X)|$ by taking the barycentric subdivision. The natural map $S(X) \to |Q(X)|$ is a homeomorphism so that the composition $S(X) \to |Q(X)| \to X$ is a weak homotopy equivalence and is a homotopy equivalence if X is homotopy equivalent to a CW complex. Then there is a natural restriction map of DGAs $\Omega^*(X) \to A_{p.l.}(S(X))$. This map induces an isomorphism on cohomology (which can be proved by induction on the cells of $|Q(X)|$). It follows that the minimal model for $\Omega^*(X)$ is identified with that of $A^*(S(X))$, uniquely up to homotopy. Hence, if X is simply connected and its rational homology is finite dimensional in each degree, then the minimal model for $\Omega^*(X)$ is dual to the rational Postnikov tower of X.

To formulate the rational differential form version of the Serre spectral sequence for a Serre fibration, it will be convenient to have a restricted type of base. Recall that if I^q is a cube, then a face of I^q is the subspace given by fixing a subset $i_1 < \cdots < i_k \leq q$ and setting $x_{i_j} = \epsilon_j$, where ϵ_j is either 0 or 1. Any face of dimension $q-k$ is identified with I^{q-k} via the complementary set of coordinates (in the same order).

Definition. A *cubical complex* X is a topological space together with set of maps of closed cubes $\sigma: I^q \to X$ with the following properties:

1. Each σ is an embedding of a closed cube into X.
2. For each closed cube $\sigma: I^q \to X$ in the set, the restriction of σ to any face of I^q is an element of the set of closed cubes in X.
3. The union of the images of the closed cubes is all of X.
4. The intersection of two closed cubes is either empty or a face of each.
5. When the intersection of two closed cubes σ_1 and σ_2 is nonempty, the induced identifications of $\sigma_1 \cap \sigma_2$ with a lower dimensional cube coming from σ_1 and σ_2 agree.

Lemma 16.3. *If* X *is a CW complex, then* X *is homotopy equivalent to a cubical complex.*

Proof. First, suppose that Δ^n is a standard n-simplex with coordinates $(t_0, \ldots, t_n)$ subject to $t_i \geq 0$ for every i and $\sum_i t_i = 1$. For $i = 0, \ldots, n$, we set C(i) equal to the subset of Δ^n where $t_i \geq t_j$ for every $j \neq i$. We define an isomorphism $I^n \xrightarrow{\cong} C(i)$ by

$$(s_1, \ldots, s_n) \mapsto \left(\frac{s_1}{1+|s|}, \ldots, \frac{s_{i-1}}{1+|s|}, \frac{1}{1+|s|}, \frac{s_i}{1+|s|}, \ldots, \frac{s_n}{1+|s|} \right),$$

where $|s| = \sum_j s_j$ and, for each j, the variable s_j ranges from 0 to 1. It is easy to see that these cubes define a cubical complex on Δ^n. Furthermore, the induced cubical complex on any face of Δ^n agrees with this construction applied to that face. Notice that the identification $I^n \to C_i$ depends on the ordering of the vertices of Δ^n.

Now suppose that K is a simplicial complex and suppose that we have a total ordering of the vertices of K. Then performing the above construction in each simplex produces a cubical complex structure on K. From this, it follows that any simplicial complex is homotopy equivalent to a cubical complex; in fact, it is piecewise linearly isomorphic to a cubical complex.

To complete the proof, we appeal to the result Lemma 2.2 that says that any CW complex is homotopy equivalent to a simplicial complex. ∎

16.2 Hirsch Extensions and Spectral Sequences

We begin by recalling the notion of a Hirsch extension. Let A^* be a connected DGA over $\mathbb{Q}$ with differential d, let n be a natural number, and let V be a finite-dimensional rational vector space. Given a map $\beta: V^* \to A^{n+1}$ whose image lies in the kernel of d, we have a *Hirsch extension in degree* n of DGA's by forming

$$A^* \to A^* \otimes_\beta \Lambda^*(V^*), \text{ with } \hat{d}|_{A^*} = d, \text{ and } \hat{d}(v) = \beta(v).$$

Here $\Lambda^*(V^*)$ is the free graded-commutative algebra generated by V^* in degree n. We denote the Hirsch extension by $\hat{A}^*$

There is a spectral sequence associated to this construction. We define a decreasing filtration on $\hat{A}^*$ by defining $F^p(\hat{A}^*) = A^{*\geq p} \otimes \Lambda^*(V^*)$. Clearly, this is a multiplicative filtration preserved by $\hat{d}$. It is easy to identify the first few terms of the spectral sequence:

$$E_0^{p,q} = A^p \otimes (\Lambda^*(V^*))^q;\ \ d_0 = 0;$$
$$E_1^{p,q} = A^p \otimes (\Lambda^*(V^*))^q;\ \ d_1 = d;$$
$$E_2^{p,q} = H^p(A^*; (\Lambda^*(V^*))^q) = H^p(A) \otimes (\Lambda^*(V^*))^q.$$

(The identification of E_2 requires that A^* be simply connected.) This is a multiplicative spectral sequence and the induced multiplication on $E_2^{*,*}$ is the (graded) tensor product of the multiplication in $H^*(A^*)$ induced by the multiplication in A^* and the usual multiplication in $\Lambda^*(V^*)$. In particular, when A^* is simply connected, multiplication induces an isomorphism

$$E_2^{p,0} \otimes E_2^{0,q} \to E_2^{p,q}.$$

The differentials $d_2, \ldots, d_n$ vanish and $d_{n+1}\colon E_{n+1}^{p,kn} \to E_{n+1}^{p+n+1,(k-1)n}$ is given by

$$H^p(A^*) \otimes (\Lambda^*(V^*))^q \to H^{p+n+1}(A^*) \otimes (\Lambda^*(V^*))^{q-n}$$

where the map is

$$a \otimes (v_1 \wedge \cdots \wedge v_k) \mapsto \sum_i (-1)^{n(i-1)} a \cup [\beta(v_i)] \otimes (v_1 \wedge \cdots \wedge v_{i-1} \wedge v_{i+1} \cdots \wedge v_k),$$

and $[\beta(v_i)] \in H^{n+1}(A^*)$ is the class represented by the closed form $\beta(v_i)$.

In [26], Sullivan showed that if $f\colon E \to B$ is a Serre fibration with fiber $K(\pi, n)$ with $\pi_1(B)$ acting trivially on the homology of the fiber, with $V = \pi \otimes \mathbb{Q}$ being finite dimensional, and if $\rho\colon A^* \to \Omega^*(B)$ is a map of DGA's inducing an isomorphism on cohomology, then there is a Hirsch extension $\hat{A}^* = (A^* \otimes \Lambda^*(V^*), \hat{d})$ and a map $\hat{\rho}\colon \hat{A}^* \to \Omega^*(E)$ with $\hat{\rho}|_{A^*} = f^* \circ \rho$ inducing an isomorphism on cohomology. The degree of the extension is n. The proof was an inductive one over the cells of the base.

Here, we propose a different argument (which also has induction over the cells in the internal parts of the argument) to prove that the Hirsch extension of the forms on the base models the total space. Our approach is to construct an DGA, denoted $\Omega^*(E/B)$, which is a sub-DGA of $\Omega^*(E)$ of cubical polynomial forms on $|Q(E)|$. This sub-DGA has on it a decreasing filtration whose E_1-term is the differential forms on the base with coefficients in the cohomology of the fiber and whose E_2-term is the cohomology of the base with coefficients in the cohomology of the fiber. If the fiber is a $K(\pi, n)$, then a DGA model for the fiber is $\Lambda^*(V^*)$ where $V = \text{Hom}(\pi, \mathbb{Q})$ and V^* is in degree n. Thus, the E_2-term of the spectral sequence for $\Omega^*(E/B)$ is isomorphic to the E_2-term of the spectral sequence of the associated Hirsch extension of $\Omega^*(B)$. It is then simply a matter of a comparison theorem for spectral sequences to prove that there is a map from the Hirsch extension $\Omega^*(B) \otimes_d \Lambda(V^*)$ to $\Omega^*(E)$ extending the natural map of $\Omega^*(B) \to \Omega^*(E)$ and inducing an isomorphism on cohomology.

To motivate the construction that we shall do in the case of Serre fibrations, let us recall briefly what one does in the case of locally trivial smooth fibrations. Suppose that $f: M \to B$ is a proper smooth submersion of smooth manifolds. Then $f: M \to B$ is a smoothly locally trivial fibration. In this case, there is a filtration on the (smooth) differential forms $\Omega^*(M)$ defined by letting $F^k(\Omega^n(M))$ be the subspace of forms ω such that for any point p in M and any collection of tangent vectors $\tau_1, \ldots, \tau_n$ with at least $n - k + 1$ of the τ_i being vertical (i.e., in the kernel of df), we have $\omega_p(\tau_1, \ldots, \tau_n) = 0$. This is a multiplicative filtration preserved by exterior differentiation, and hence, there is an induced spectral sequence whose $E_1^{p,q}$-term is identified with $\Omega^p(B; H^q(F; \mathbb{R}))$ where the coefficients are the local system given by the cohomology of the fibers of the map. This is a locally trivial coefficient system over B and d_1 is identified with d_B, the usual differential for this locally trivial system, so that $E_2^{p,q}$ is identified with the cohomology of B with coefficients in this local system, $H^p(B; H^q(F; \mathbb{R}))$. In the case when $\pi_1(B)$ acts trivially on the cohomology of the fiber, the coefficients become a trivial local system identified with $H^q(F_0; \mathbb{R})$, for the fiber F_0 over the base point. Then $E_2^{p,q} = H^p(B; H^q(F_0; \mathbb{R})) = H^p(B; \mathbb{R}) \otimes_{\mathbb{R}} H^q(F_0; \mathbb{R})$. In this case, the multiplication of forms induces a product structure on the spectral sequence which agrees with the obvious product on $E_2^{*,*}$.

16.3 Polynomial Forms for a Serre Fibration

Now suppose that $f: E \to B$ is a Serre fibration with the base B being a cubical complex. (Recall that this means that each of the closed cubes is embedded in B.) Our goal is to give a $\mathbb{Q}$-DGA model $f^*: \Omega^*(B) \to \Omega^*(E/B)$ for $f: E \to B$ with a filtration on $\Omega^*(E/B)$ analogous to the one above giving the Serre spectral sequence (with rational coefficients) for this Serre fibration. To do this, we define sets of cubes associated to this fibration, $C_*(E/B)$. An element of $C_k(E/B)$ is a singular cube $\sigma: I^k \to E$ for which there is a closed cube $\bar{\sigma}$ of B, with identification $\varphi_{\bar{\sigma}}: I^s \to \bar{\sigma} \subset B$ such that setting p equal to the projection $p: I^k \to I^s$ given by $p(x_1, \ldots, x_k) = (x_1, \ldots, x_s)$, we have $f \circ \sigma = \varphi_{\bar{\sigma}} \circ p$. In this case, we say that σ *covers* $\bar{\sigma}$. Notice that σ covers only one cube of B. We denote by $D_k(E/B)$ the set of degenerate cubes in $C_k(E/B)$.

Suppose that $\sigma \in C_k(E/B)$ covers $\bar{\sigma}$ of dimension s. Then for $\epsilon = 0, 1$ for any $i > s$, the face $f_i^\epsilon \sigma$ also covers $\bar{\sigma}$ and for $i \le s$ the face $f_i^\epsilon \sigma$ covers $f_i^\epsilon \bar{\sigma}$. It follows that if f is a face of σ, then f covers a face $\bar{f}$ of $\bar{\sigma}$ and the dimension of the fiber of $f \to \bar{f}$ is less than or equal to the dimension of the fiber of $\sigma \to \bar{\sigma}$. We let $Q_*(E/B)$ denote the chain complex whose chain group in degree k is the quotient of the free abelian group generated by $C_k(E/B)$ by the free abelian group generated by $D_k(E/B)$ and whose boundary map is the one induced by taking sums in the standard way of codimension-1 faces. The subcomplex $D_*(E/B)$ is closed under taking boundaries; also if singular cube covering $\bar{\sigma}$ is degenerate, then its nondegenerate quotient also covers $\bar{\sigma}$. Clearly, $C(E/B)$ is a subset of $C(E)$, and the resulting maps of chain groups embed $Q_*(E/B)$ as a subcomplex of $Q_*(E)$. Ideas from [21] can be used to show the following:

Proposition 16.4. *The inclusion of* $Q_*(E/B) \to Q_*(E)$ *induces an isomorphism on homology so that* $Q_*(E/B)$ *computes the singular homology of* E.

Proof. We first prove the result for bases, B that are finite cubical complexes. The proof is by induction on the number of cubes in the cubical complex structure of B. If B has a single cell, then it is a point and $Q_*(E/B)$ is just the usual singular cubical complex of E which by [21] or Lemma 16.1 computes the singular homology of E. Now suppose by induction that we know the result for all bases with fewer than N cells and B has exactly N cells. Let $B_0 \subset B$ be the sub-cubical complex of B obtained by removing the interior one of the top-dimensional cubes, say $\bar{\sigma}$ of dimension s. Let $E_0 = f^{-1}(B_0)$. Then $E_0 \to B_0$ is a Serre fibration and by induction $Q_*(E_0/B_0)$ computes the singular homology of E_0. Clearly, $Q_*(E_0/B_0)$ is a subcomplex of $Q_*(E/B)$ so that we have a commutative diagram of short exact sequences of chain complexes whose chain groups are free abelian groups:

$$\begin{array}{ccccccccc} 0 \to & Q_*(E_0/B_0) & \to & Q_*(E/B) & \to & Q_*(E/B)/Q_*(E_0/B_0) & \to 0 \\ & \downarrow & & \downarrow & & \downarrow & \\ 0 \to & Q_*(E_0) & \to & Q_*(E) & \to & Q_*(E, E_0) & \to 0 \end{array}$$

where the vertical maps are the inclusions. By induction and the 5 lemma to establish that the middle vertical arrow induces an isomorphism on homology, it suffices to show that the last vertical arrow does. Let F_0 be the fiber over the initial vertex v of $\bar{\sigma}$. By the fact that singular cubical theory computes the usual singular homology, we have $H_*(Q_*(E, E_0)) = H_*(\bar{\sigma}, \partial\bar{\sigma}) \otimes H_*(F_0)$.

Now we define a map

$$K_{\bar{\sigma}}: C_*(F_0) \to C_{s+*}(f^{-1}(\bar{\sigma})/\bar{\sigma}), \tag{16.1}$$

denoted $K_{\bar{\sigma}}(\tau) = \hat{\tau}$, by induction satisfying the following properties:

1. For $\tau: I^k \to F_0$, the map $\hat{\tau}: I^s \times I^k \to f^{-1}(\bar{\sigma})$ satisfies $\hat{\tau}|_{\{v\}\times I^k} = \tau$.
2. If τ is degenerate with τ_0 as nondegenerate quotient, then $\hat{\tau}$ is degenerate with nondegenerate quotient $\hat{\tau}_0$.
3. For each face $f_i^\epsilon = f_i^\epsilon(\tau)$ of τ, we have $\hat{\tau}|_{I^s \times f_i^\epsilon} = \widehat{f_i^\epsilon(\tau)}$.

Suppose that for each singular cube $\tau: I^t \to F_0$ of dimension $t < k$, we have defined $\hat{\tau}: I^s \times I^k \to E_0$ covering $\bar{\sigma}$ with these properties. We consider a singular cube τ of dimension k. If τ is degenerate, with nondegenerate quotient τ_0, the second condition defines $\hat{\tau}$ from $\hat{\tau}_0$. Suppose that τ is nondegenerate. By induction, we have $\widehat{\partial\tau}: I^s \times \partial I^k \to E_0$ covering $\partial\bar{\sigma}$. We also have $\tau: \{v\} \times I^k \to F_0 \subset E_0$. We extend this to a map $\hat{\tau}_1: I^1 \times I^k$ covering $\bar{\sigma}|I^1$ and compatible with the restriction of $\widehat{\partial\tau}$ to $I^1 \times \partial I^k$. Such an extension is possible by the Serre homotopy extension property and the fact that $I^1 \times I^k$ deformation retracts onto $\{v\} \times I^k \cup I^1 \times \partial I^k$. We continue inductively in this way extending τ to $\hat{\tau}_i$ over $I^i \times I^k$ covering the restriction of $\bar{\sigma}$ to I^i and compatible with the given extension on the boundary. This uses the Serre homotopy extension property and the fact that for each $i \leq s$ the cube $I^i \times I^k$ deformation retracts onto $I^{i-1} \times I^k \cup I^i \times \partial I^k$. In the end, we construct $\hat{\tau}$ as required. We do this independently for each nondegenerate k cube, completing the induction.

This map determines a chain map

$$K_{\bar{\sigma}}: Q_*(F_0) \to Q_{s+*}(f^{-1}(\bar{\sigma})/\bar{\sigma})/Q_*(f^{-1}(\partial\bar{\sigma})/\partial\bar{\sigma})) = Q_*(E/B)/Q_*(E_0/B_0)$$

commuting with the boundary maps.

Claim. $K_{\bar{\sigma}}$ induces an isomorphism

$$H_*(F_0) \longrightarrow H_{s+*}(Q_*(E/B), Q_*(E_0/B_0)).$$

Proof. Given an element $\tau: I^s \times I^k$ covering $\bar{\sigma}$, then the restriction of τ to $\{v\} \times I^k$ is an element τ_v of $C_k(F_0)$ and hence $K_{\bar{\sigma}}(\tau_v)$ is an element of $C_{k+s}(E/B)$ covering $\bar{\sigma}$. Notice that this map is compatible with taking faces in the last k-directions. This defines a chain map $L_{\bar{\sigma}}: Q_{s+k}(E/B)/Q_{s+k}(E_0/B_0)$ to $C_s(F_0)$. Clearly, $L_{\bar{\sigma}} \circ K_{\bar{\sigma}} = Id_{Q_*(F_0)}$. We complete the proof by showing that $K_{\bar{\sigma}} \circ L_{\bar{\sigma}}$ is chain homotopic to the identity of $Q_*(E/B)/Q_*(E_0/B_0)$. To do this, we define a map $H_{\bar{\sigma}}: Q_*(E, E_0) \to Q_{*+1}(E, E_0)$ compatible with taking faces in directions greater than s and also compatible with degeneracies and satisfies

$$\partial H_{\bar{\sigma}} - H_{\bar{\sigma}} \circ \partial = Id - K_{\bar{\sigma}} \circ L_{\bar{\sigma}}.$$

This chain homotopy is defined inductively in the same way as the map $K_{\bar{\sigma}}$, above. This completes the proof of the claim. ∎

It follows immediately from the claim that the third vertical arrow in the above diagram induces an isomorphism on homology. The inductive step follows immediately from this and the 5 lemma. This shows that provided that B is a finite cubical cell complex, the result holds.

Now any cubic complex is a direct limit of its finite subcomplexes and homology commutes with direct limits, so the result follows in general from the result for bases that are finite cell complexes. This completes the proof of the proposition. ∎

Corollary 16.5. *Integration defines a map of cochain complexes* $\Omega^*(E/B) \to Q^*(E; \mathbb{Q})$ *that induces an isomorphism on cohomology.*

Proof. Since the inclusion $Q_*(E/B) \to Q_*(E)$ induces an isomorphism on homology, the result is immediate from Proposition 16.2. ∎

16.4 Serre Spectral Sequence for Polynomial Forms

Now we introduce the decreasing filtration on $\Omega^*(E/B)$ and compute the E_1- and E_2-terms in the resulting spectral sequence. A filtration analogous to the one in the smooth case makes sense on $\Omega^*(E/B)$. We say that a form $\{\omega_\sigma\}_\sigma$ in $\Omega^n(E/B)$ is contained in $F^k(\Omega^n(E/B))$ if for any singular cube σ in E covering a cube $\bar{\sigma}$ of dimension s of the cubical structure on B and tangent vectors $(v_1, \ldots, v_n)$ to I^n at a point $a \in I^n$ the evaluation $\omega_\sigma(v_1, \ldots, v_n)$ is zero any time, at least $n - k + 1$ of the vectors v_i are tangent to the fiber of the projection $p: I^n \to I^s$, given by $p(x_1, \ldots, x_n) = (x_1, \ldots, x_s)$. Notice that since the cubes in E are required to cover embedded cubes in B, the filtration is preserved as we pass to faces: If ω_σ is in filtration level p, then the same is true for the restriction of ω_σ to any face. This is a multiplicative filtration preserved by d. Hence, it induces a spectral sequence with a multiplication.

For each $\sigma \in C_*(E/B)$ covering a cube $\bar{\sigma}$ of the cubical structure of B, we denote by $p_\sigma: I^n \to \bar{\sigma}$ the natural projection and by F_σ the fiber $p_\sigma^{-1}(v_{\bar{\sigma}})$ where $v_{\bar{\sigma}}$ is the initial vertex of $\bar{\sigma}$. Here is the main theorem of this appendix.

Theorem 16.6. *Suppose that* $f: E \to B$ *is a Serre fibration with* B *being a path connected cubical complex. Let* $b \in B$ *be a base point and let* $F_0 \subset E$ *be the fiber over* b. *Suppose that* $\pi_1(B, b)$ *acts trivially on the homology of* F_0. *Then the filtration on* $\Omega^*(E/B)$ *defined above induces a multiplicative spectral sequence whose* E_2*-term is given by*

$$E_2^{p,q} = H^p(B; \mathbb{Q}) \otimes H^q(F_0; \mathbb{Q}),$$

and with the multiplication on E_2*—being the (graded) tensor product of the cup product structures on the cohomology of the base and the fiber. In particular,* $E_2^{p,0} \otimes E_2^{0,q} \to E_2^{p,q}$ *is an isomorphism.*

Proof. We begin with a series of claims. ∎

Claim. $F^k(\Omega^*(E/B))/F^{k+1}(\Omega^*(E/B))$ is naturally identified with collections of elements $\{\omega_\sigma\}_\sigma$ indexed by $\sigma \in C_*(E/B)$ with $\omega_\sigma \in p_{\bar{\sigma}}^*\Omega^k(\bar{\sigma}) \otimes \Omega^*(F_{\bar{\sigma}}) \subset \Omega^*(\sigma)$ and compatible under the face relations and degeneracies (the degeneracies being only along the fiber direction). The map induced by exterior differentiation on F^k/F^{k+1} is given by $(-1)^k \otimes d_{F_{\bar{\sigma}}}$ on $p_{\bar{\sigma}}^*\Omega^k(\bar{\sigma}) \otimes \Omega^*(F_{\bar{\sigma}})$.

Proof. Let $\bar{\sigma}$ be the cube of B covered by σ. Then $F^k(\Omega^*(\sigma/\bar{\sigma}))$ is identified with $p_{\bar{\sigma}}^*\Omega^{*\geq k}(\bar{\sigma}) \otimes \Omega^*(F_{\bar{\sigma}})$, and we have an identification

$$F^k(\Omega^*(\sigma/\bar{\sigma}))/F^{k+1}(\Omega^*(\sigma/\bar{\sigma})) = p_{\bar{\sigma}}^*\Omega^k(\bar{\sigma}) \otimes \Omega^*(F_{\bar{\sigma}}).$$

The face relations preserve the bi-degrees as do the degeneracies. The latter are only in the fiber direction since $\bar{\sigma}$ is nondegenerate. This proves that any element in the associated graded determines a collection of elements as stated in the claim.

Conversely, given a compatible collection of elements, they define an element in $\Omega^*(E/B)$ which lies in F^k and the linear subspace of such elements is a complement for $F^{k+1} \subset F^k$. This proves the first statement in the claim. The second is clear: On $\Omega^*(\sigma) = p_{\bar{\sigma}}^*\Omega^*(\bar{\sigma}) \otimes \Omega^*(F_{\bar{\sigma}})$, exterior differentiation is given by $d = d \otimes 1 + (-1)^k 1 \otimes d$. The first term raises the filtration degree by 1 the second preserves it; thus, on the associated graded, the first term vanishes. ■

Now consider $\Omega^*(B)$; it is the DGA of homogeneous collections $\{\omega_{\bar{\sigma}}\}_{\bar{\sigma}}$ with $\omega_{\bar{\sigma}} \in \Omega^*(\bar{\sigma})$ indexed by the cells $\bar{\sigma}$ of B and compatible under restriction to faces. (Since we use only the cells of B and not all singular cubes in B, there are no degeneracies to consider.) This DGA has an integration mapping to $Q^*(B; \mathbb{Q})$ which induces an isomorphism on cohomology. Of course, $\Omega^*(B)$ is a quotient DGA of the $\Omega^*(|Q(B)|)$ dual to the cellular inclusion $B \hookrightarrow |Q(B)|$, and hence, this projection of DGA's induces an isomorphism on cohomology.

At this point, we choose a base point $b \in B$ (say the initial 0-cube of B under some ordering) and we let F_0 be the fiber over b. We suppose that $\pi_1(B, b)$ acts trivially on the homology of F_0. Given any $b' \in B$, choosing a path from b to b' identifies the homology or cohomology of F_0 with that of the fiber F' over b'. Because of the assumption that $\pi_1(B, b)$ acts trivially on the homology of F_0, this identification is independent of the path chosen to connect b to b'.

The next step is to define a map of $E_1^{k,*}$ to $\Omega^k(B) \otimes H^*(F_0; \mathbb{Q})$ sending the multiplication in E_1 to the wedge product of forms tensored with the cup product in cohomology. Let $\{\omega_\sigma\}_\sigma \in p_{\bar{\sigma}}^*\Omega^k(\bar{\sigma}) \otimes \Omega^*(F_{\bar{\sigma}})$ be a compatible collection of forms determining an element in $E_0^{k,*}$. Suppose further that this element is in the kernel of d_0, meaning that for each σ the form ω_σ lies in $p_{\bar{\sigma}}^*\Omega^k(\bar{\sigma}) \otimes Z^*(F_{\bar{\sigma}})$ where $Z^* \subset \Omega^*$ is the kernel of d. Such an element determines a class in $E_1^{k,*}$ and every class in $E_1^{k,*}$ is determined by such elements. Lastly, two such elements determine the same class in $E_1^{k,*}$ if and only if they differ by a collection of forms $\{d\eta_\sigma + \beta_\sigma\}$ where $\{\eta_\sigma\}_\sigma$ is a compatible collection of forms in $\Omega^k(\bar{\sigma}) \otimes \Omega^{*-1}(F_{\bar{\sigma}})$ and $\{\beta_\sigma\}_\sigma$ is a compatible collection of forms in $\Omega^{k+1}(\bar{\sigma}) \otimes \Omega^*(F_{\bar{\sigma}})$.

Given a form $\{\omega_\sigma\}_\sigma$ with the property that $\omega_\sigma \in p_{\bar{\sigma}}^*\Omega^k(\bar{\sigma}) \otimes Z^*(F_{\bar{\sigma}})$ for every σ, then we restrict attention to the ω_σ indexed by the σ covering a given cell $\bar{\sigma}$ of B. Denote the set of such σ by $C_*(\bar{\sigma})$. In this way, we are restricting to $\Omega^*(f^{-1}(\bar{\sigma})/\bar{\sigma})$. We pull $\{\omega_\sigma\}_{C_*(\bar{\sigma})}$ back via $K_{\bar{\sigma}}$ to give a class in $\Omega^k(\bar{\sigma}) \otimes Z^*(F_{\bar{\sigma}})$. Projecting modulo $\Omega^k(\bar{\sigma}) \otimes d\Omega^{*-1}(F_{\bar{\sigma}})$ determines an element in $\Omega^k(\bar{\sigma}) \otimes H^*(F_{\bar{\sigma}})$.

Claim. For any choice of $K_{\bar{\sigma}}$ as in Equation 16.1, the resulting map

$$K_{\bar{\sigma}}^*: \Omega^*(f^{-1}(\bar{\sigma})/\bar{\sigma}) \to \Omega^*(\bar{\sigma}) \otimes \Omega^*(F_{\bar{\sigma}})$$

has the following properties:

1. The push forward of the filtration F^* via $K_{\bar{\sigma}}^*$ is the filtration which is the product of the filtration by degrees on $\Omega^*(\bar{\sigma})$ and the trivial filtration on $\Omega^*(F_{\bar{\sigma}})$.
2. The $E_1^{p,q}$-term of the spectral sequence for $F^*(\Omega^*(\bar{\sigma}) \otimes \Omega^*(F_{\bar{\sigma}}))$ is naturally identified with $\Omega^p(\bar{\sigma}) \otimes H^q(F_{\bar{\sigma}}; \mathbb{Q})$.

3. The pullback by $K^*_{\bar{\sigma}}$ induces an isomorphism on the $E_1^{*,*}$ terms on the spectral sequences associated to $F^*(\Omega^*(f^{-1}(\bar{\sigma})/\bar{\sigma}))$ and $F^*(\Omega^*(\bar{\sigma}) \otimes \Omega^*(F_{\bar{\sigma}}))$, so that the $E_1^{p,q}$-term of the spectral sequence of $F^*(\Omega^*(f^{-1}(\bar{\sigma}))$ is also naturally identified with $\Omega^p(\bar{\sigma}) \otimes H^q(F_{\bar{\sigma}}; \mathbb{Q})$.

Proof. Clearly, the induced filtration on $\mathrm{Im}K_{\bar{\sigma}}$ agrees with the filtration on $\Omega^*(\bar{\sigma}) \otimes \Omega^*(F_{\bar{\sigma}})$ which is given by the degree in the first factor. Using this filtration, both the map $K^*_{\bar{\sigma}}: \Omega^*(f^{-1}(\bar{\sigma})) \to \Omega^*(\bar{\sigma}) \otimes \Omega^*(F_{\bar{\sigma}})$ and the map $r^*_{\bar{\sigma}}: \Omega^*(\bar{\sigma}) \otimes \Omega^*(F_{\bar{\sigma}}) \to \Omega^*(f^{-1}(\bar{\sigma}))$ are filtration preserving as is the homotopy map $H^*_{\bar{\sigma}}: \Omega^*(f^{-1}(\bar{\sigma})/\bar{\sigma}) \to \Omega^*(f^{-1}(\bar{\sigma})/\bar{\sigma}) \otimes \Lambda(s, ds)$. Thus, $\int_{s=0}^{s=1} H^*_{\bar{\sigma}}$ forms a filtration-preserving cochain homotopy from Id to $r^*_{\bar{\sigma}} \circ K^*_{\bar{\sigma}}$. It then follows that $K^*_{\bar{\sigma}}$ and $r^*_{\bar{\sigma}}$ are inverse isomorphisms on the $E_1^{*,*}$-terms of the spectral sequences associated to these filtrations. Clearly for $\Omega^*(\bar{\sigma}) \otimes \Omega^*(F_{\bar{\sigma}})$ with the filtration coming from the degree of the first factor, the $E_1^{p,q}$-term is naturally identified with $\Omega^p(\bar{\sigma}) \otimes H^q(\Omega^*(F_{\bar{\sigma}}))$. ■

Now we wish to replace $H^*(F_{\bar{\sigma}}); \mathbb{Q})$ in the above lemma by $H^*(F_0; \mathbb{Q})$,where F_0 is the fiber over the base point. To do this, we need only remark that choosing any path from v_0 to $v_{\bar{\sigma}}$ identifies the cohomology of $F_{\bar{\sigma}}$ with that of F_0 and this identification is independent of the choice of path. Thus, for each cell $\bar{\sigma}$ of B, we have an identification of $E_1^{p,q}(F^*(\Omega^*(f^{-1}(\bar{\sigma})/\bar{\sigma})))$ with $\Omega^p(\bar{\sigma}) \otimes H^q(F_0; \mathbb{Q})$.

Next, we need to see that these identifications are compatible under passing from $\bar{\sigma}$ to a face $\bar{\tau}$. First, notice that the maps of the $E_1^{*,*}$-terms determined by two product structures $K_{\bar{\sigma}}$ and $K'_{\bar{\sigma}}$ are equal since they are both inverses to the map induced by $r_{\bar{\sigma}}$ on the $E_1^{*,*}$-terms. Next, notice that the map induced by a product structure over $\bar{\sigma}$ from $H^*(F_{\bar{\sigma}}; \mathbb{Q})$ to the rational cohomology of the fiber over any other vertex v' of $\bar{\sigma}$ agrees with the isomorphism induced by taking a path from $v_{\bar{\sigma}}$ to v' and hence commutes with the identifications of the cohomologies of these fibers with that of F_0. Using these two facts, the result is immediate. Hence, we have constructed a map from $E_1^{p,q}(\Omega^*(E/B)$ to $\Omega^p(B) \otimes H^q(F_0; \mathbb{Q})$.

The last thing to see is that this map is an isomorphism. This requires a claim:

Claim. The restriction mapping

$$\Omega^*(f^{-1}(\bar{\sigma})/\bar{\sigma}) \to \Omega^*(f^{-1}(\partial\bar{\sigma})/\partial\bar{\sigma})$$

induces a surjective map on filtration level p for every $p \geq 0$.

Proof. Suppose by induction, we have extended a form

$$\omega_\sigma \in F^p(\Omega^*(f^{-1}(\partial\bar{\sigma})/\partial\bar{\sigma}))$$

to a form in filtration level p defined on all cubes covering $\bar{\sigma}$ of dimension less than k, and let σ be a cube covering $\bar{\sigma}$ of dimension k. (By the inductive hypothesis, the form is defined on $\partial\sigma$ and is in filtration level p.) The extension result requires inductively taking linear combinations of forms defined on opposite boundaries. This does not change the filtration level and hence shows that we can extend the form over all of σ keeping the result in filtration level p. ■

Now we are ready to show that $E_1^{p,q}(\Omega^*(E/B)) = \Omega^p(B) \otimes H^q(F_0; \mathbb{Q})$. The proof is first for the case when B is a finite cell complex and is by induction on the number of cells of B, the case of one cell being obvious. Suppose we know the result when the base has fewer than N cells and B has N cells. Let B_0 be the result of removing a top-dimensional cell of B. By induction, $E_1^{p,q}(\Omega^*(E_0/B_0)) = \Omega^p(B_0) \otimes H^q(F_0; \mathbb{Q})$. Given any element $\alpha \in \Omega^p(B) \otimes H^q(F_0; \mathbb{Q})$, we fix an element $q_0 \in F^p(\Omega^*(E_0/B_0))$ which is in the kernel of d_0 and represents the restriction α_0 of α in $E_1^{p,q}(\Omega^*(E_0/B_0))$. Choose closed forms $\zeta_1, \dots, \zeta_n, \dots$, in $\Omega^q(F_{\bar{\sigma}})$ whose cohomology classes form a basis for $H^q(F_{\bar{\sigma}}; \mathbb{Q})$. Then the restriction of α to $E_1^{p,q}(\Omega^*(f^{-1}(\partial\bar{\sigma})/\partial\bar{\sigma}))$ is represented as a finite sum $\sum_{i=1}^{K} \omega^i \otimes \zeta_i$ with $\omega^i \in \Omega^p(\partial\bar{\sigma})$. Let $\hat{\omega}^i \in \Omega^p(\bar{\sigma})$ be an extension of ω^i. Then the element $\mu = \sum_i \hat{\omega}^i \otimes \zeta_i$ is an element in $F^p(\Omega^*(f^{-1}(\bar{\sigma})/\bar{\sigma}))$ which represents an element in $E_1^{p,q}(\Omega^*(f^{-1}(\bar{\sigma})/\bar{\sigma}))$ whose restriction to the boundary agrees in E_1 with the restriction of α_0 to $f^{-1}(\partial\bar{\sigma})$. Thus, there difference of these restriction is contained in $F^{p+1} + dF^p$ of $\Omega^*(f^{-1}(\partial\bar{\sigma})/\partial\bar{\sigma})$. Using the surjective property established in the previous claim, we can vary μ by a form in $(F^{p+1} + dF^p)(\Omega^*(f^{-1}(\bar{\sigma})/\bar{\sigma}))$ to arrange that its restriction agrees with the restriction of α_0 in $\Omega^*(f^{-1}(\partial\bar{\sigma})/\partial\bar{\sigma})$. This then gives an extension of the form over all of B and proves that the map $E_1^{p,q} \to \Omega^p(B) \otimes H^q(F_0; \mathbb{Q})$ is surjective.

Now we must show that it is injective. Again, this is done inductively. Suppose that $\alpha \in F^p(\Omega^*(E/B))$ is in the kernel of d_0 and maps to the trivial element in $\Omega^p(B) \otimes H^q(F_0; \mathbb{Q})$. By induction, the restriction, α_0, of α in $F^p(\Omega^*(E_0/B_0))$ is contained in $F^{p+1} + dF^p$. Using the surjectivity result above, we can assume that $\alpha_0 = 0$. This means that α is the extension by zero over E_0 of a relative form α_1 in $\Omega^*(f^{-1}(\bar{\sigma})/\bar{\sigma}, f^{-1}(\partial\bar{\sigma})/\partial\bar{\sigma})$ in filtration level p. The image of α_1 in E_1 is contained in $\Omega^p(\bar{\sigma}, \partial\bar{\sigma}) \otimes H^q(F_0; \mathbb{Q})$ and α is contained in the image of $(F^{p+1} + dF^p)(\Omega^*(f^{-1}(\bar{\sigma})/\bar{\sigma}))$. We write $\alpha = \beta + d\gamma$ where $\beta \in F^{p+1}$ and $\gamma \in F^p$. Since $\alpha_!$ restricts to zero over $\partial\bar{\sigma}$, we have

$$(\beta + d\gamma)|_{f^{-1}(\partial\bar{\sigma})} = 0.$$

This means that $\gamma|_{f^{-1}(\partial\bar{\sigma})}$ represents in element in $E_1^{p,q-1}(\partial\bar{\sigma})$. By the surjectivity result, there is $\hat{\gamma} \in F^p(\Omega^*(f^{-1}(\bar{\sigma}))$ with the property that $\hat{\gamma}|_{f^{-1}(\partial\bar{\sigma})} = \gamma|_{\partial\bar{\sigma}}$. Varying α by $d\hat{\gamma}$ allows us to assume that $\gamma|_{f^{-1}(\partial\bar{\sigma})} = 0$ and hence that also that$\beta|_{f^{-1}(\partial\bar{\sigma})} = 0$. Thus, $\alpha_1 = \beta + d\gamma$ in $\Omega^*(f^{-1}(\bar{\sigma}), f^{-1}(\partial\bar{\sigma}))$. Thus, α is equal to the extension of $\beta + d\gamma$ by 0 outside of $f^{-1}(\bar{\sigma})$. This shows that α is an element in $F^{p+1} + dF^p$ and hence represents 0 in $E_1^{p,q}$.

This completes the proof when the base is a finite complex. When the base is an infinite complex, $\Omega^*(E/B)$ and $F^p(\Omega^*(E/B)$ are the projective limits over finite subcomplexes $B_n \subset B$ of the $\Omega^*(f^{-1}(B_n)/B_n)$ and $F^p(\Omega^*(f^{-1}(B_n)/B_n))$. Since the restriction mappings are surjective on all filtration levels, it follows that the E_0-term commutes with the projective limits. Since $E_1^{p,q}(f^{-1}(B_n)/B_n)$ is identified with $\Omega^p(B_n) \otimes H^q(F_0; \mathbb{Q})$, it follows that the restriction mappings in the projective system are surjective on the cohomology of (E_0, d_0). It follows that the $E_1^{p,q}$ commutes with the projective limits and hence $E_1^{p,q}(\Omega^*(E/B)) = \Omega^p(B) \otimes H^q(F_0; \mathbb{Q})$. Clearly, the map d_1 is given by $d_B \otimes 1$, so that the $E_2^{p,q}(\Omega^*(E/B)) = H^p(B; H^q(F_0; \mathbb{Q}))$.

16.5 Proof of Theorem 12.1

Suppose that B is a cubical complex and that $f\colon E \to B$ is a Serre fibration with fiber $K(\pi, n)$ with $V = \pi \otimes \mathbb{Q}$ a finite-dimensional rational vector space and with the action of $\pi_1(B, b)$ on the homology of the fiber being trivial. Then V^* is the relative $(n+1)^{st}$ cohomology associated with the map $f^*\colon \Omega^*(B) \to \Omega^*(E/B)$. That is to say, there is a map

$$V^* \longrightarrow \Omega^{n+1}(B) \oplus \Omega^n(E/B)$$

whose image lies in the relative cocycles and which induces the identity on the relative cohomology in degree $n+1$. This defines a map

$$\rho\colon \Omega^*(B) \otimes_d \Lambda^*(V^*) \to \Omega^*(E/B).$$

We have the decreasing filtration on $\Omega^*(B) \otimes_d \Lambda^*(V^*)$ coming from the degrees of forms in $\Omega^*(B)$. With this filtration and the filtration described above on $\Omega^*(E/B)$, the map ρ preserves the filtration. Given the isomorphisms of $E_2^{p,0}$ of each spectral sequence with $H^p(B)$, the induced map on $E_2^{p,0}$ is the identity. Also, the $E_2^{0,n}$-term of each spectral sequence is identified with the relative $(n+1)^{st}$ cohomology, and again the map between them is the identity. It then follows from the multiplicative structure that the induced map at E_2 is an isomorphism, and hence, the map on cohomology is an isomorphism.

Now suppose that $\mathfrak{M} \to \Omega^*(B)$ is a minimal model. Let $\mathfrak{M}' = \mathfrak{M} \otimes_d \Lambda(V^*)$ be the Hirsch extension corresponding to $f\colon E \to B$. Then $\mathfrak{M}'$ is the minimal model for $\Omega^*(B) \otimes_d \Lambda^*(V^*)$ and hence for $\Omega^*(E/B)$ and hence also for $\Omega^*(E)$.

Lastly, suppose that $f'\colon E' \to B'$ is a simplicial model for $f\colon E \to B$. Then $\mathfrak{M}$ is also a minimal model for the p.l. forms $A^*(B)$. Then the minimal model for the p.l. forms $A^*(E')$ is identified with the minimal model for $\Omega^*(E)$, and hence, the map $\mathfrak{M}' \to \Omega^*(E)$ determines a map $\mathfrak{M}' \to A^*(E')$ up to homotopy, and this map induces an isomorphism on cohomology. The restriction of this map to $\mathfrak{M}$ factors as

$$\mathfrak{M} \xrightarrow{\rho} A^*(B') \xrightarrow{(f')^*} A^*(E'),$$

where ρ induces an isomorphism on cohomology. Also, the relative $(n+1)^{st}$ cohomologies of both $(\mathfrak{M}, \mathfrak{M}')$ and of $(A^*(B'), A^*(E'))$ are identified with V^*. Under these identifications, ρ induces the identity on these relative cohomology groups. This proves that given a principal $K(\pi, n)$-fibration $E \to B$ and a simplicial model $E' \to B'$ with minimal model $\rho\colon \mathfrak{M} \longrightarrow A^*(B')$, the Hirsch extension $\mathfrak{M}' = \mathfrak{M} \otimes_d \Lambda^*(V^*)$ dual to the principal fibration maps to $A^*(E')$ extending the composition $(f')^* \circ \rho$ and inducing an isomorphism on cohomology. This is exactly the statement of Theorem 12.1.

Chapter 17
Quillen's Work on Rational Homotopy Theory

Before Sullivan's work on differential forms and rational homotopy theory, in [19] Quillen had established algebraic models for rational homotopy theory. Quillen worked dually from the way Sullivan does: Instead of using differential forms as the basic model, Quillen uses differential graded Lie algebras. One can, as Quillen does, associate a differential graded, co-commutative co-algebra to a differential graded Lie algebra. Dualizing this produces a differential graded algebra computing the rational homotopy type and thus in some sense solves the rational commutative cochain problem. Sullivan's construction was much more directly tied to differential forms on the space, whereas Quillen's went through more homotopy theoretic constructions, i.e., the simplicial set (of singular simplices) associated with the loop space of the given space.

In this chapter we will briefly describe Quillen's results and compare them with Sullivan's. We begin by describing the homotopy theory of various categories of algebraic objects that enter into Quillen's construction.

17.1 Differential Graded Lie Algebras

Definition. A *graded Lie algebra (GLA)* is a graded vector space $L = \oplus_k L^k$ together with a homomorphism

$$[\cdot,\cdot]\colon L \otimes L \to L$$

satisfying:

1. $[\cdot,\cdot]$ is homogeneous of degree zero in the sense that $[\cdot,\cdot]\colon L^k \otimes L^n \to L^{k+n}$.
2. (Graded skew symmetry): For homogeneous elements x, y, we have $[x, y] = (-1)^{|x|\cdot|y|+1}[y, x]$.
3. (Graded Bianchi identity): For homogeneous elements x, y, z, we have

P. Griffiths and J. Morgan, *Rational Homotopy Theory and Differential Forms*, Progress in Mathematics 16, DOI 10.1007/978-1-4614-8468-4_17,
© Springer Science+Business Media New York 2013

$$[[x, y], z] + (-1)^{|z|(|x|+|y|)}[[z, x], y] + (-1)^{|x|(|z|+|y|)}[[y, z], x] = 0.$$

(Here, for a homogeneous element $x \in L^k$, the symbol $|x|$ is defined to be k.)

A *differential graded Lie algebra (DGLA)* is a graded Lie algebra $(L, [\cdot, \cdot])$ together with a differential $\partial: L \to L$, which is a linear map of degree -1 satisfying $\partial^2 = 0$ and which satisfies

$$\partial([x, y]) = [\partial x, y] + (-1)^{|x|}[x, \partial y]$$

for homogeneous elements x, y. The *homology* of a DGLA is the homology of (L, ∂). Clearly, the homology of a DGLA is a GLA. The DGLAs that come up in Quillen's construction are *reduced*, meaning that $L^k = 0$ for $k \leq 0$. A *map (morphism)* of DGLAs is a degree-preserving linear map that preserves the bracket and commutes with the differentials. A *quasi-isomorphism (or homotopy equivalence)* is a map that induces an isomorphism on homology. The *homotopy category of DGLA* is the category whose objects are DGLAs and whose morphisms are compositions of DGLA maps and formal inverses of quasi-isomorphisms.

In passing from spaces to DGLAs, there is a shift in degree by -1 from the usual notion of degree in a space to the degree in the DGLA, so that reduced DGLAs are associated to simply connected spaces. Suppose that X is a 1-connected space with base point x and that $(L(X), \partial)$ is the DGLA associated to it by Quillen's construction (as will be described below). Let ΩX denote the based loop space of X. Then, the homology $H_k(L(X), \partial)$ is identified with the rational homotopy group $\pi_{k+1}(X) \otimes \mathbb{Q} = \pi_k(\Omega X) \otimes \mathbb{Q}$. Furthermore, the induced graded Lie algebra structure on $H_*(L(X), \partial)$ is identified with the Whitehead product on rational homotopy groups, which recall is a pairing $\pi_k(X) \otimes \pi_n(X) \to \pi_{k+n-1}(X)$ and is graded skew symmetric after shifting down one degree. This product is identified with the product $\pi_{k-1}(\Omega X) \otimes \pi_{n-1}(\Omega X) \to \pi_{k+n-2}(\Omega X)$ induced by the loop composition product $\Omega X \times \Omega X \to \Omega X$.

17.2 Differential Graded Co-algebras

Co-algebras are not as familiar as algebras, but they are the dual objects to algebras, meaning that all the properties that are familiar for algebras have dual analogues for co-algebras. The dual properties are formulated by simply reversing the directions of all arrows in the defining diagrams for the property for algebras.

Definition. We fix a ground field K and all vector spaces are K-vector spaces. A *co-algebra* is a vector space C with a co-multiplication

$$\Delta: C \to C \otimes C.$$

The co-algebra is said to be *co-associative* if

$$(\Delta \otimes \mathrm{Id}_C) \circ \Delta = (\mathrm{Id}_C \otimes \Delta) \circ \Delta$$

as maps of C to $C \otimes C \otimes C$. A *graded co-algebra* is a graded vector space $C = \oplus_k C_k$ with a co-multiplication that is homogeneous of degree zero, meaning that for each k

$$\Delta : C_k \to \oplus_{i+j=k} C_i \otimes C_j.$$

A co-algebra is *(graded) co-commutative* if $\Delta(c) = T \circ \Delta$ where $T : C \otimes C \to C \otimes C$ is the map that sends $c_1 \otimes c_2 \mapsto (-1)^{|c_1| \cdot |c_2|} c_2 \otimes c_1$ for all homogeneous elements c_1, c_2. We always assume that our co-algebras are connected, meaning that $C_i = 0$ for $i < 0$ and C_0 is the ground field and co-unital, meaning that part of the structure is a co-unit $\epsilon : C \to K$ so that $\epsilon \otimes \mathrm{Id}_C \circ \Delta$ is the natural map $C \to K \otimes_K C$ and analogously for $\mathrm{Id} \otimes \epsilon \circ \Delta$. In the case of connected co-algebras ϵ identifies C_0 with K and the co-unital condition can be rewritten as

$$\Delta(c) = c \otimes 1 + 1 \otimes c + \overline{\Delta}(c)$$

with $\overline{\Delta}(c) \in \sum_{i,j>0} C_i \otimes C_j$.

A *differential graded co-algebra* is a graded co-algebra together with a boundary map $\partial : C \to C$ of degree -1 satisfying $\partial^2 = 0$ and satisfying the co-Leibnitz rule: Suppose that $\Delta(c) = \sum_i c'_i \otimes c''_{k-i}$ where each c'_i is homogeneous. Then,

$$\Delta(\partial c) = \sum_i \partial c'_i \otimes c''_i + (-1)^{|c'_i|} c'_i \otimes \partial c''_i.$$

A *DGCC* means a differential graded co-commutative co-associative co-algebra.

A *map (morphism)* of DGCCs is a degree-preserving co-algebra map. A morphism is a *quasi-isomorphism (homotopy equivalence)* if it induces an isomorphism on the homology. The *homotopy category of DGCCs* has as objects DGCC and as morphism compositions of morphisms of DGCCs and formal inverses of quasi-isomorphisms.

17.3 The Bar Construction

There are many variants of the bar construction. The one that is relevant for us here goes from a differential graded Lie algebra to a differential graded co-commutative, co-associative co-algebra.

Definition. Let V_* be a graded vector space. A co-commutative, co-associative graded co-algebra C_* is *co-free with* V_* *as co-generating vector space* if it satisfies the following universal property. There is a degree zero linear map $\pi : C_* \to V_*$

(degree zero meaning that it preserves degrees), and given any co-commutative, co-associative graded co-algebra C'_* with a degree zero linear map $f: C'_* \to V_*$, there is a unique degree zero co-algebra map $\hat{f}: C'_* \to C_*$ with $\pi \circ \hat{f} = f$.

Let us give a construction of a co-free algebra with V_* as co-generating vector space. We form the subspace of graded symmetric elements $S^*(V_*)$ inside the tensor algebra by taking the sub-vector space of elements in the tensor algebra $T^*(V_*)$ that are invariant under the signed action of the symmetric group:

$$\sigma: V_* \otimes \cdots \otimes V_* \to V_* \otimes \cdots \otimes V_*$$

acts by $\sigma(v_1 \otimes \cdot \otimes v_n) = \epsilon v_{\sigma(1)} \otimes \cdots \otimes v_{\sigma(n)}$ where the sign $\epsilon = \epsilon(v_1 \otimes \cdots \otimes v_n)$ is the Kozul sign which is the sign homomorphism generated by assigning $(-1)^{|v_i| \cdot |v_{i+1}|}$ to the interchange permutation of i and $i + 1$. This determines $S^*(V_*)$. (Notice that it is not a subalgebra of the tensor algebra.)

Next we introduce a co-multiplication which will make $S^*(V_*)$ co-free in the category of co-commutative, co-associative graded co-algebras. To do this, we introduce the shuffle coproduct:

$$\Delta(\sigma(v_1) \otimes \cdots \otimes \sigma(v_k)) = \sum_{(\mu,\nu)} \epsilon(v, \mu, \nu) v_\mu \otimes v_\nu.$$

Let us explain the notation. Here (μ, ν) ranges over all shuffles of $\{1, \ldots, k\}$, meaning that $\mu = \mu_1 < \mu_2 < \cdots < \mu_i$ and $\nu = \nu_1 < \cdots < \nu_j$, and $\{\mu_1, \ldots, \mu_i, \nu_1, \ldots, \nu_j\}$ is a permutation of $\{1, \ldots, k\}$. Furthermore, $v_\mu = v_{\mu_1} \otimes \cdots \otimes v_{\mu_i}$ and similarly for v_ν. The sign $\epsilon(v, \mu, \nu) \in \{\pm 1\}$ is the Kozul sign associated with the permutation and the degrees of the v_i. To see that $S^*(V_*)$ is co-free with V_* as co-generating space, suppose that C'_* is a graded co-commutative co-algebra and $f: C'_* \to V_*$ is a degree zero linear map. Then, we define $f_k: C_* \to S^k(V_*)$ by

$$f_k = \underbrace{(f \otimes \cdots \otimes f)}_{k-\text{times}}(\eta^{k-2} \circ \eta^{k-3} \circ \cdots \circ \eta^0)$$

where

$$\eta^r = \underbrace{1 \otimes \cdots \otimes 1}_{r-\text{times}} \otimes \overline{\Delta}.$$

One sees directly that, provided that C'_* is co-commutative and co-associative, the f^k together define the required co-algebra map from $C'_* \to S^*(V_*)$.

Now we apply this construction beginning with a DGLA $(L = \oplus_k L_k, \partial)$. We shift up a degree by forming $V = \Sigma L$, where by definition $(\Sigma L)_k = L_{k-1}$. Then, we form the co-algebra $S^*(\Sigma L)$ with the shuffle coproduct. For any $a \in L_k$, we denote by $\sigma(a)$ the corresponding element in $(\Sigma L)_{k+1}$. This produces a co-commutative, co-free co-algebra with ΣL as co-free co-generating space.

So far we have not made use of two of the data from the DGLA: the Lie algebra product and the differential $\partial: L \to L$. It turns out that each of these can be used to define a differential on $S^*(\Sigma L)$ and that these differentials commute. The differential $\partial: L \to L$ defines a differential, denoted $\partial^{int}: S^*(\Sigma L) \to S^*(\Sigma L)$, by requiring that $\partial^{int}(\sigma(a)) = -\sigma\partial(a)$ for all $a \in L$ and by requiring that ∂ satisfy the co-Leibnitz rule. It follows immediately that $(\partial^{int})^2 = 0$. The Lie bracket defines a differential ∂^{ext} by requiring $\partial^{ext}(\sigma(a) \otimes \sigma(b) + (-1)^{|a|\cdot|b|}\sigma(b) \otimes \sigma(a)) = (-1)^{|a|}\sigma[a, b]$. The Jacobi identity implies that $(\partial^{ext})^2 = 0$, and the fact that ∂ commutes with the Lie bracket implies that $\partial^{int}\partial^{ext} + \partial^{ext}\partial^{int} = 0$. We then define the differential $\hat{\partial} = \partial^{int} + \partial^{ext}$.

The resulting co-associative, co-commutative, (co-free) differential graded co-algebra $(S^*(\Sigma L), \hat{\partial})$ is the *bar construction* applied to (L, ∂). This is a functor from the category of DGLAs to the category of DGCCs and it sends quasi-isomorphisms to quasi-isomorphisms.

17.4 Relationship Between Quillen's Construction and Sullivan's

Before giving Quillen's construction of a DGLA associated to a simply connected space, let us state the relationship of his construction to Sullivan's.

One thing to point out is that Sullivan's construction assumes that the rational homotopy groups are finite dimensional in all dimensions so that passing from the rational homotopy groups to their duals and dualizing again reproduces the rational homotopy groups. Since Quillen works with co-algebras, he does not have to dualize and hence does not need to make this assumption. Apart from that proviso, the constructions are dual in the following sense:

Theorem 17.1. *Quillen assigns a reduced DGLA,* $(L(X), \partial)$, *to a simply connected topological space* X *in a functorial way. Applying the bar construction to* $(L(X), \partial)$ *produces a co-commutative differential graded co-algebra* $Q_*(X)$. *Dualizing this produces a DGA* $Q^*(X)$ *functorially associated to* X. *Suppose that the rational homotopy groups of* X *in each degree are finite dimensional. If* K *is a simplicial complex and* $f: K \to X$ *is a weak homotopy equivalence, then the minimal model* $\mathfrak{M}_K$ *for the p.l. forms* $A^*(K)$ *and the minimal model* $\mathfrak{M}_X$ *for the DGA* $Q^*(X)$ *are identified by a map determined up to homotopy by the homotopy class of* f.

17.5 Quillen's Construction

We have not yet described how Quillen associates a DGLA to a simply connected topological space X. To explain this requires more algebro-topological machinery.

Simplicial categories. Quillen's construction uses simplicial objects, so we begin with a brief introduction to them. There is a category Δ with objects the ordered sets $\Delta^n = \{0, \ldots, n\}$, $n = 0, 1, \ldots$. The set Δ^n is called the n-*simplex*. The morphism set

$$\mathrm{Hom}_\Delta(\Delta^k, \Delta^n)$$

is the set of all set functions f from $\{0, \ldots, k\}$ to $\{0, \ldots, n\}$ that are weakly order-preserving, in the sense that if $a \leq b$ then $f(a) \leq f(b)$. Composition of morphisms in the category is given by the usual composition of set functions. These morphisms can also be viewed as simplicial maps from the geometric k-simplex to the geometric n-simplex that weakly preserve the order of the vertices.

Given a category $\mathcal{C}$, the *simplicial* $\mathcal{C}$ *category* has as its objects' covariant functors from the opposite category to Δ, denoted Δ^{opp}, to $\mathcal{C}$ and as its morphisms natural transformations between functors. In particular, an object $\mathcal{O}$ in simplicial $\mathcal{C}$ consists of an indexed family $\{\mathcal{O}_n\}_{n\geq 0}$ of objects of $\mathcal{C}$. For each order-preserving simplicial map $\alpha: \Delta^k \to \Delta^n$, there is a dual morphism $\alpha^*: \mathcal{O}_n \to \mathcal{O}_k$ of $\mathcal{C}$ and compositions (with order reversed) are preserved. There is an especially nice generating set of morphisms: For each n, there are n+1 face maps from $\Delta^{n-1} \to \Delta^n$ denoted $f_0, \ldots, f_n$. In terms of simplicial maps, f_i is the inclusion of Δ^{n-1} as the codimension-1 face of Δ^n opposite the ith vertex of Δ^n. In terms of set functions $f_i(a) = a$ if $a < i$ and $f_i(a) = a + 1$ if $a \geq i$. Dually, there are *n* degeneracy maps $\Delta^n \to \Delta^{n-1}$, denoted $s_0, \ldots, s_{n-1}$, where s_i is the projection of Δ^n to its codimension-1 face opposite the ith vertex. In terms of set maps, $s_i(a) = a$ for $a \leq i$ and $s_i(a) = a - 1$ for a i. For each object $\mathcal{O}$ in $\mathcal{C}$ there morphisms which are the images of these. The image of f_i is a "boundary map" $\partial_i: \mathcal{O}_n \to \mathcal{O}_{n-1}$ and the image of s_i is a degeneracy map $s_i: \mathcal{O}_{n-1} \to \mathcal{O}_n$. The relations in Δ lead to the following relations:

$$\partial_i \circ \partial_j = \partial_{j-1} \circ \partial_i \quad \text{if } j \geq i + 1.$$

$$s_i \circ s_j = s_{j+1} \circ s_i \quad \text{if } j \geq i.$$

$$\partial_i \circ s_j = s_{j-1} \circ \partial_i \quad \text{if } j \geq i + 1.$$

$$\partial_i \circ s_j = s_j \circ \partial_{i-1} \quad \text{if } j + 1 < i.$$

$$\partial_i \circ s_i = \partial_{i+1} \circ s_i = \mathrm{Id}.$$

All morphisms of the category are compositions of these generating morphisms, and two compositions of these generators give the same morphism of the category if there is a sequence of compositions connecting them with each step in the sequence being one of these relations (pre- and post composed by the same pair of compositions).

Some examples are in order:

1. A *simplicial set* K has as objects indexed families of sets $\{K_n\}_{n\geq 0}$ and for each n set functions $\partial_i: K_n \to K_{n-1}$ for $0 \leq i \leq n$ and set functions $s_i: K_{n-1} \to K_n$ for $0 \leq i < n$ satisfying the relations given above. A morphism from a simplicial set K to a simplicial set K′ consists of set functions $\varphi_n: K_n \to K'_n$ commuting with the boundary and degeneracy maps. A simplicial set K with a base point $x_0 \in K_0$ has homotopy groups, which are easiest to define when the simplicial set is a Kan complex, meaning that it satisfies the following extension property: Given $0 \leq k \leq n$ and elements $a_0, \ldots, a_{k-1}, a_{k+1}, \ldots, a_n$ in K_{n-1} with $\partial_i a_j = \partial_{j-1} a_i$ for all $i < j$ with $i \neq k$ and $j \neq k$, then there is an element $a \in K_n$ with $\partial_i a = a_i$ for all $i \neq k$. (See below for the definition of the homotopy groups of a simplicial set.)
2. The category of simplicial spaces has as objects indexed families of topological spaces $\{X_n\}_{n\geq 0}$ and continuous maps $\partial_i: X_n \to X_{n-1}$ and $s_i: X_{n-1} \to X_n$ satisfying the relations given above. Morphisms are indexed families of continuous maps commuting with the boundary and degeneracy maps.
3. There are simplicial categories associated with various algebraic categories, e.g., simplicial groups, simplicial (augmented and completed) Hopf algebras, and simplicial DGLAs.
4. The first and motivating example is that of the simplicial set of a topological space X. The simplicial set associated to X has $\mathcal{S}_n(X)$ equal to the set of singular n-simplices of X, i.e., the set of continuous maps $\Delta^n \to X$. The boundaries, ∂_i, are given by restricting to the various codimension-1 faces. More generally, given any morphism $\varphi: \Delta^k \to \Delta^n$ of Δ, we can view φ as a simplicial map from k-simplex to the n-simplex. Precomposing with this map induces a map from $\mathcal{S}_n(X) \to \mathcal{S}_k(X)$. Notice that $\mathcal{S}(X)$ is a Kan complex. The notions of a simplicial set and of the simplicial set associated with a topological space were introduced by Kan in [11].

As we remarked above, the homotopy groups of a simplicial set K are easiest to define when the simplicial set satisfies is a Kan complex. For a Kan complex K with a base point $x_0 \in X_0$, we define the homotopy groups as follows. Consider all $x \in K_n$ with the property that $\partial_i x = x_0^{n-1} := s_{n-2}s_{n-3}\cdots s_0 x_0$ for all $0 \leq i \leq n$. We say that two such, $x, y \in K_n$, are equivalent if there is $z \in K_{n+1}$ with $\partial_n z = x$, $\partial_{n+1} z = y$ and with $\partial_i z = x_0^n$ for all $0 \leq i < n$. Because of the Kan condition, this is an equivalence relation. We denote the set of equivalence classes by $\pi_n(K, x_0)$. To define the multiplication of two elements $x, y \in \pi_n(K)$, consider $a_0 = \cdots = a_{n-2} = x_0^n, a_{n-1} = x, a_{n+1} = y$. This is a compatible family of elements of K_n, so that by the Kan condition, there is $z \in K_{n+1}$ with $\partial_i z = a_i$ for all $i \neq n$. We define xy to be the equivalence class of $\partial_n z$. This is a well-defined group operation on $\pi_n(K, x_0)$, which is commutative if $n > 1$. In general, to define the homotopy groups, one has to replace a simplicial set by an equivalent Kan complex, which can always be done.

Given a simplicial set, one can construct its *geometric realization*. This is a simplicial complex whose n-simplicies are the nondegenerate n-simplicies in the simplicial set and whose gluing maps are determined by the face maps in the simplicial set. If X is a topological space and $|\mathcal{S}(X)|$ is the geometric realization of the singular simplicial set of X, then there is a natural continuous map $|\mathcal{S}(X)| \to X$ which induces an isomorphism on the homotopy groups and the homology groups. We can then take the rational differential forms on this simplicial complex and hence form Sullivan's minimal model construction for any topological space, by replacing the space by the geometric realization of its singular simplicial set.

If X is a simplicial complex, then ordering the vertices of X identifies it with a subcomplex of $\mathcal{S}(X)$ and hence there is a natural restriction map $A^*(|\mathcal{S}(X)|) \to A^*(X)$, a map that induces an isomorphism on cohomology, and hence these DGAs are homotopy equivalent and in particular the minimal models for these two DGAs of p.l. forms are identified, uniquely up to homotopy.

From a simply connected based space to its 1-trivial simplicial set. The first step in Quillen's construction uses a variant of the simplicial set of a space, a variant that can be defined for simply connected spaces. Let X be a path connected, simply connected space with a base point x_0 and form a simplicial set $\{E^i_{2,\mathrm{Sing}}(X)\}_{i\geq 0}$ where $E^i_{2,\mathrm{Sing}}(X)$ is the set of all continuous maps of the i-simplex into X with the property that the one-skeleton of Δ^i maps to the base point x_0. This is an object in the category of 1-trivial simplicial sets K, i.e., those with the property that the set K^i is a single point for $i = 0, 1$.

From a 1-trivial simplicial set to its reduced simplicial loop group. One then replaces a 1-trivial simplicial set K, for example, $K = E_{2,\mathrm{Sing}}(X)$, by a simplicial group G(K) that is the simplicial analogue of the based loop space $\Omega(X, x_0)$ when $K = E_{2,\mathrm{Sing}}(X)$. The group $G(K)_n$ is the free group generated by the set of all elements of K_{n+1} that are not contained in $s_n(K_n)$. For $0 \leq i < n$, the boundary map $\bar{\partial}_i: G(K)_n \to G(K)_{n-1}$ is induced by $\partial_i: K_{n+1} \to K_n$ and $\bar{\partial}_n x = \partial_n x \partial_{n+1}(x)^{-1}$. For $0 \leq i \leq n$, the degeneracy map $\bar{s}_i: G(K)_n \to G(K)_{n+1}$ is induced by the degeneracy $s_i: K_{n+1} \to K_{n+2}$. Then, G(K) is an object in the category of reduced simplicial groups, i.e., those with $G_0 = \{e\}$. Any simplicial group is a Kan complex, and therefore, we can directly define the homotopy groups. But there is another more direct description of the homotopy groups of a simplicial group. We define the normalized chains N(G) of a simplicial group G to be the complex

$$N_q(G) = \cap_{0<j\leq q} \mathrm{Ker}\,\partial_j: G_q \to G_{q-1}.$$

The map ∂_0 induces a homomorphism $N_q(G) \to N_{q-1}(G)$ whose composition with itself is trivial. The homotopy groups of the simplicial group G are given by

$$\pi_q(G, e) = \frac{\mathrm{Ker}\,\partial_0: N_q \to N_{q-1}}{\mathrm{Im}\,\partial_0(N_{q+1})}.$$

For all $q \geq 1$ these are abelian groups. When $X_1 = X_0$ is a single point, then $G_0(X)$ is the trivial group and hence the homotopy groups of G(X) are all abelian in this case. We define a rational equivalence between simplicial groups G, G' with

$G_0 = G'_0 = \{e\}$ to be a morphism of simplicial groups inducing an isomorphism on the rational homotopy groups.

At this point we have associated to a simply connected space a connected simplicial group, the one associated to the 1-trivial simplicial set of singular simplices in X. The next step in Quillen's construction is a passage to Hopf algebras, which needs an introductory description.

Hopf Algebras. There is one more algebraic category that comes up in Quillen's construction—that of Hopf algebras. Suppose that A is an associative algebra with unit, then $A \otimes A$ is becomes an algebra with the multiplication determined by $(a_1 \otimes a_2) \otimes (b_1 \otimes b_2) = a_1 b_1 \otimes b_2 b_2$. A *Hopf algebra* over $\mathbb{Q}$ is an algebra R over $\mathbb{Q}$ together with maps

$$\Delta : R \to R \otimes R, \quad \eta : R \to \mathbb{Q}, \quad \tau : R \to R \text{ denoted } x \mapsto \bar{x},$$

satisfying:

1. Δ is a co-associative, co-commutative co-multiplication which is algebra homomorphism.
2. η is a (two-sided) co-unit.
3. $x \mapsto \bar{x}$ is an antipode, meaning that it is an anti-homomorphism, and if $\Delta(h) = \sum_i h'_i \otimes h''_i$, then

$$\sum_i \bar{h'_i} \cdot h''_i = \sum_i h'_i \cdot \bar{h''_i} = \eta(h)1.$$

A map (morphism) of Hopf algebras is a linear map preserving the multiplication, co-multiplication, unit and co-unit, and the antipode.

Recall that a *shuffle* of $(1, \ldots, k)$ is a division $(i_1, \ldots, i_r), (j_1, \ldots, j_s)$ where $i_1, \ldots, i_r, j_1, \ldots, j_s$ is a partition of $1, \ldots, k$ and also $i_1 < \cdots < i_r$ and $j_1 < \cdots < j_s$. Associated to a Lie algebra we have its Universal Enveloping Algebra which is a Hopf algebra, though the algebra structure is not commutative. The algebra structure is the quotient of the tensor algebra on L, $T^*(L) = \oplus_{k \geq 0} L^{\otimes k}$, by the relation $xy - yx = [x, y]$. for $x, y \in L$. The co-multiplication Δ is given by the shuffle coproduct:

$$\Delta(x_1 \otimes \cdots \otimes x_k) = \sum_{a,b} x_a \otimes x_b$$

where (a, b) runs over the shuffles of $1, \ldots, n$, and if $a = (i_1, \cdots, i_r)$, then $x_a = x_{i_1} \otimes \cdots \otimes x_{i_r}$. This is the unique co-multiplication that is a homomorphism of algebras sending x to $x \otimes 1 + 1 \otimes x$ for $x \in L$. It is a co-commutative co-multiplication. The antipode of $x_1 \cdots x_k$ is defined to be $(-1)^k x_k \cdots x_1$. The Poincaré–Birkhoff–Witt theorem says that L embeds in its Universal Enveloping Algebra as the space of primitive elements (those satisfying $\Delta(a) = a \otimes 1 + 1 \otimes a$). In fact, there is an equivalence of categories between Lie algebras and Hopf algebras.

A nice feature of this map is captured in the co-multiplication. It turns out that the co-commutative co-algebra underlying the Universal Enveloping Hopf algebra is in fact co-free in the category of co-commutative co-algebras. Said another way this co-algebra is isomorphic to the co-algebra on the graded symmetric algebra generated by L shifted up one degree with the co-multiplication being given by the shuffle coproduct as above. We define the *augmentation ideal* $\bar{R}$ of a Hopf algebra R to be the kernel of η.

Quillen works with a related category—the category of reduced Hopf algebras, completed with respect to the augmentation ideal. Given a vector space V with a decreasing filtration $V = F^0(V) \supset F^1(V) \supset \cdots$, we have the associated graded vector space

$$Gr(V) = \oplus_{n \geq 0} Gr^n(V) = \oplus_{n \geq 0} F^n(V)/F^{n+1}(V).$$

Suppose that R is an augmented algebra (augmented, meaning that there is a homomorphisms $\epsilon: R \to \mathbb{Q}$). We denote by $\bar{R}$ the ideal that is the kernel of ϵ. We consider filtrations $F^*(R)$ with the property that $F^0 = R$ and $F^1(R) = \bar{R}$ and $F^i(R) = \text{Im} \otimes_i \bar{R} \to R$. This filtration is multiplicative in the sense that multiplication induces a map $F^i(R) \otimes F^j(R) \to F^{i+j}(R)$. Then, Gr(R) is a graded algebra, and we can form completion of R with respect to the augmentation ideal by setting $\hat{R} = \varprojlim (R/F^n(R))$. It satisfies:

1. $F^1(\hat{R})$ is the augmentation ideal of $\hat{R}$.
2. $Gr(\hat{R})$ is generated by $Gr^1(\hat{R})$
3. $\hat{R} = \varprojlim (\hat{R}/F^n(\hat{R}))$.

In this case there is a completed tensor product $\hat{R} \hat{\otimes} \hat{R}$ defined by completing $\hat{R} \otimes \hat{R}$ with respect to the filtration $F^k(\hat{R} \otimes \hat{R}) = \oplus_{i+j=k} F^i(\hat{R}) \otimes F^j(\hat{R})$. A completed Hopf algebra is an algebra $\hat{R}$, complete with respect to the augmentation ideal, with a map

$$\Delta: \hat{R} \to \hat{R} \hat{\otimes} \hat{R}$$

which is a co-associate, co-commutative co-multiplication map of complete algebras.

The primary example for us comes from the rational group ring $\mathbb{Q}G$ of a group G. The multiplication operation in G defines an algebra structure on $\mathbb{Q}G$. There is also a co-commutative, co-associative co-multiplication

$$\Delta: \mathbb{Q}G \to \mathbb{Q}G \otimes \mathbb{Q}G$$

defined by $\Delta(g) = g \otimes g$ for every $g \in G$. This, together with the natural maps $\eta: \mathbb{Q}G \to \mathbb{Q}$ defined by the homomorphism $G \to \{e\}$ and $\tau: \mathbb{Q}G \to \mathbb{Q}$ defined by $\tau(g) = g^{-1}$, makes $\mathbb{Q}G$ a Hopf algebra. We complete this Hopf algebra with respect to the augmentation ideal to define $\widehat{\mathbb{Q}G}$, a complete Hopf algebra. An element x of a Hopf algebra is *primitive* if

$$\Delta(x) = x \otimes 1 + 1 \otimes x,$$

any primitive element of a complete Hopf algebra is contained in the augmentation ideal. An element x is *group-like* if $\Delta(x) = x \otimes x$. Group-like elements in a complete Hopf algebra are congruent to 1 modulo in the augmentation ideal. For any element x in the augmentation ideal of $\widehat{\mathbb{Q}G}$, the power series exp(x) is an element of $\widehat{\mathbb{Q}G}$. Furthermore, if x is primitive, then the power series exp(x) is a group-like. Conversely, if x is a group-like element, then the power series log(x) is a primitive element. These are inverse identifications between the sets of group-like and primitive elements.

Now we are ready to complete our description of Quillen's construction. To a topological space X, we associated the simplicial group G(X) which is the analogue of the loop space applied to the simplicial set of (1-trivial) singular simplices in X. Here are the remaining steps.

From a reduced simplicial group to its reduced simplicial Hopf algebra. Quillen passes from the category of reduced simplicial groups to the category of reduced simplicial Hopf algebras over $\mathbb{Q}$ that are complete with respect to the augmentation ideal. This passage is the application of the simplicial functor to the map that associates to a group the completion $\widehat{\mathbb{Q}G}$ of the rational group ring $\mathbb{Q}G$ with respect to the augmentation ideal. Since $G_0 = \{e\}$, the simplicial Hopf algebra is reduced in the sense that its degree zero Hopf algebra is $\mathbb{Q}$. The map $g \to g \otimes g$ for all $g \in G$ extends to a co-multiplication on $\mathbb{Q}G$, making it a Hopf algebra and makes $\widehat{\mathbb{Q}G}$ a complete Hopf algebra. There is an adjoint to this functor that associates to a reduced, complete Hopf algebra its group of group-like elements. Applying the simplicial functor produces adjoint functors between the category of simplicial groups and the category of reduced simplicial complete Hopf algebras. A morphism between reduced simplicial complete Hopf algebras over $\mathbb{Q}$ is said to be a rational equivalence if the induced map of simplicial groups of group-like elements induces an isomorphism on rational homotopy groups.

From a reduced simplicial Hopf algebra to its connected simplicial Lie algebra. We can associate to a reduced simplicial complete Hopf algebra the simplicial Lie algebra of primitive elements of the Hopf algebra. This is a reduced simplicial Lie algebra in the sense that $L_0 = 0$. A morphism between two reduced simplicial Lie algebras is a rational equivalence if the induced map on homotopy groups (which are $\mathbb{Q}$-vector spaces) is an isomorphism.

From a connected simplicial Lie algebra to its connected DGLA. Lastly, we pass from a simplicial Lie algebra to a differential graded Lie algebra as follows. Given a simplicial vector space V, we define the *normalized chains* of V by

$$N_q(V) = \cap_{0<j\leq q}\mathrm{Ker}(\partial_j\colon V_q \to V_{q-1})$$

with the boundary map given by

$$\partial = \partial_0\colon N_q \to N_{q-1}.$$

In the case when L is a simplicial Lie algebra, we can define the structure of a differential graded Lie algebra on $N_*(L)$ by using the Eilenberg–Zilber map. Suppose that V and W are simplicial vector spaces and let $V \otimes W$ be the graded vector space whose degree q subspace is $V_q \otimes W_q$ for each $q \geq 0$. For $x \in V_p$ and $y \in W_q$, we define the Eilenberg–Zilber map $x \underline{\otimes} y \in (V \otimes W)_{p+q}$ by the formula

$$x \underline{\otimes} y = \sum_{(\mu,\nu)} \operatorname{sign}(\mu,\nu) s_{\nu_q} \cdots s_{\nu_1} x \otimes s_{\mu_p} \cdots s_{\mu_1} y,$$

where (μ,ν) runs over all (p,q)-shuffles of $\{0,\ldots,p+q-1\}$; i.e.,

$$(\mu,\nu) = (\mu_1,\ldots,\mu_p,\nu_1,\ldots,\nu_q)$$

is a permutation with $\mu_1 < \cdots < \mu_p$ and $\nu_1 < \cdots < \nu_q$, and $\operatorname{sign}(\mu,\nu)$ is the sign of the permutation.

We define the bracket operation $N(L)_p \otimes N(L)_q \to N(L)_{p+q}$ by mapping $x \otimes y$ to the image under the bracket map $L_{p+q} \otimes L_{p+q} \to L_{p+q}$ of the image $x \underline{\otimes} y$ under the Eilenberg–Zilber map. Again a morphism is a rational equivalence if it induces an isomorphism on the homology of the DGLAs.

This then is Quillen's functorial assignment of a DGLA to a simply connected topological space. As we have already remarked, we can apply the bar construction to pass from a DGLA to a co-commutative, co-associative co-algebra and then by duality to a DGA. The resulting DGA has the same minimal model as the p.l. forms on a simplicial complex homotopy equivalent to X.

Chapter 18
A_∞-Structures and C_∞-Structures

Before Quillen's theory of the homotopy theory of various algebraic categories including differential Lie algebras and differential co-algebras were introduced and before Sullivan's theory of the homotopy theory of DGAs were introduced, there were developments in algebraic topology and homotopy theory that, while it was not fully appreciated at the time, are, in hindsight, closely related to these theories. These developments began with A_∞-spaces and A_∞-structures in the 1960s and have been developed in the intervening years to what now goes under the name of the theory of operads. One of the first results of this theory was Stasheff's characterization [24] of those topological spaces that are homotopy equivalent to loop spaces as being A_∞-spaces. (Indeed, an A_∞-structure on a space determines the homotopy type of its classifying space, that is to say, its de-looping.) These ideas, originating in homotopy theory, have also been crucial recently for developments in symplectic geometry, in particular in formulating the symplectic side of the mirror symmetry conjecture which originated in physics. One now formulates the appropriate object associated to a symplectic manifold as an A_∞-category of Lagrangian submanifolds of the symplectic manifold, *the Fukaya category*.

18.1 Operads, Rooted Trees, and Stasheff's Associahedron

Operads encode in a quite general context the idea of a family of composable operations. Each operation has a number of inputs (which, in our rending of the theory, are ordered) and a single output. The set of n-ary operations (those with exactly n inputs) is denoted $\mathcal{P}_n$. Given $\mathrm{a_n} \in \mathcal{P}_\mathrm{n}$ and inputs $\mathrm{x_1, \ldots, x_n}$, the value of the operation is denoted $\mathrm{a_n(x_1, \ldots, x_n)}$. Suppose that we have operations $\mathrm{a_n} \in \mathcal{P}_\mathrm{n}$ and $\mathrm{a_j} \in \mathcal{P}_\mathrm{j}$ and suppose we fix i with $1 \leq \mathrm{i} \leq \mathrm{n}$. Then, we can compose $\mathrm{a_n}$ with $\mathrm{a_j}$ inserted at the ith-position, producing an element in $\mathcal{P}_{\mathrm{n+j-1}}$ as follows. Given inputs $\mathrm{x_1, \ldots, x_{n+j-1}}$, we apply $\mathrm{a_n}$ to the inputs

P. Griffiths and J. Morgan, *Rational Homotopy Theory and Differential Forms*, Progress in Mathematics 16, DOI 10.1007/978-1-4614-8468-4_18,
© Springer Science+Business Media New York 2013

$$x_1, \ldots, x_{i-1}, a_j(x_i, \ldots, x_{i+j-1}), x_{i+j}, \ldots, x_{n+j-1}.$$

These composition laws are formalized as follows.

An *(nonsymmetric) operad* is a collection of sets $\{\mathcal{P}_n\}_{n\geq 1}$ together with functions, defined for each $(n, j) \geq 1$ and i with $1 \leq i \leq n$,

$$\mathcal{O}[(n, j); i] \colon \mathcal{P}_j \times \mathcal{P}_n \to \mathcal{P}_{n+j-1}$$

and also a distinguished element $1 \in \mathcal{P}_1$ that acts as a unit for the operations $\mathcal{O}[(n, 1); i]$ for every $1 \leq i \leq n$. These operations are required to satisfy the following natural composition axioms:

1. When $i' < i$, the maps

$$\mathcal{O}[(n+j-1, \ell); i'] \circ \left(\mathrm{Id}_{\mathcal{P}_\ell} \times \mathcal{O}[(n, j); i]\right) : \mathcal{P}_\ell \times \mathcal{P}_j \times \mathcal{P}_n \to \mathcal{P}_{n+\ell+j-2}$$

   and

$$\mathcal{O}[(n+\ell-1, j); i+\ell-1] \circ \left(\mathrm{Id}_{\mathcal{P}_j} \times \mathcal{O}[(n, \ell); i']\right) : \mathcal{P}_j \times \mathcal{P}_\ell \times \mathcal{P}_n \to \mathcal{P}_{n+j+\ell-2}$$

   agree (after composing the map that interchanges the first two factors in the domain of the second map).
2. Similarly, when $i \leq i' \leq i+j-1$, the maps

$$\mathcal{O}[(n+j-1, \ell); i'] \circ \left(\mathrm{Id}_{\mathcal{P}_\ell} \times \mathcal{O}[(n, j); i]\right) : \mathcal{P}_\ell \times \mathcal{P}_j \times \mathcal{P}_n \to \mathcal{P}_{n+j+\ell-2}$$

   and

$$\mathcal{O}[(n, j+\ell-1); i] \circ \left(\mathcal{O}[(j, \ell); i'-i+1] \times \mathrm{Id}_{\mathcal{P}_n}\right) : \mathcal{P}_\ell \times \mathcal{P}_j \times \mathcal{P}_n \to \mathcal{P}_{n+j+\ell-2}$$

   agree.

   One can use these binary operations to define functions

$$\mathcal{P}_{\ell_1} \times \cdots \mathcal{P}_{\ell_n} \times \mathcal{P}_n \to \mathcal{P}_{\sum_{i=1}^n \ell_i}.$$

This represents the operation obtained by applying any n-ary operation to the outputs of n other operations.

A prototypical example of operads is the operad defined by rooted planar trees. $\mathcal{P}_n$ consists of the set of isomorphism classes planar trees, rooted at the bottom with leaves labelled in order left to right $1, \ldots, n$. The operation $\mathcal{O}[(n, j); i]$ is to glue the root of the rooted planar tree with j leaves to the ith leaf of the rooted planar tree with n leaves and then renumber the leaves left to right.

Given a set X, there is a natural operad *End*(X). The n-ary operations are all set functions from $X^n \to X$, and the structure maps $\mathcal{O}[(n, j); i]$ are the natural maps: for $f_n \in End(X)_n$ and $f_j \in End(X)_j$, we have

$$\mathcal{O}[(n,j);i](f_j,f_n)(x_1,\ldots,x_{n+j-1}) =$$
$$= f_n(x_1.\ldots,x_{i-1},f_j(x_i,\ldots,x_{i+j-1}),x_{i+j},\ldots,x_{n+j-1}).$$

Definition. Given an operad $\mathcal{P} = \{\mathcal{P}_n\}$, the set X is an *algebra over the operad* if there is a morphism of operads from $\mathcal{P} \to End(X)$. Said another way, for each $p_n \in \mathcal{P}_n$ determines a function $f(p_n): X^n \to X$ in such a way that under this correspondence the structure maps of the operad $\mathcal{P}$ are mapped to the corresponding compositions of functions.

Stasheff's Associahedron. This is a beautiful cell complex naturally associated to the operad of rooted planar trees. It is a cell complex whose cells describe the ways to associate strings on n-elements, i.e., intersperse matched parentheses in a meaningful way in a string of *n* elements. These parentheses are non-redundant except that they always include parentheses around the whole expression. For example, for 4 elements *abcd*, we have

$$(((ab)c)d),\ ((a(bc))d),\ (a((bc)d)),\ (a(b(cd))),\ ((ab)(cd)),$$
$$((ab)cd),\ (a(bc)d),\ (ab(cd)),\ ((abc)d),\ (a(bcd))$$
$$(abcd),$$

where the first row consists of complete associations and the second row consists of elements with two pairs of parentheses.

The dimension of the cell corresponding to an insertion of parentheses is n − 1 minus the number of pairs of parentheses. Suppose that a cell c corresponds to a set of matched parentheses p_c. Then, the face of a cell is the union of cells corresponding to sets of parentheses $p_{c'}$ that include p_c as a subset (with the same matching), i.e., all meaningful sets of parentheses $p_{c'}$ that can be obtained from p_c by adding more pairs of matched parentheses.

The associahedron for 4 elements is the pentagon:

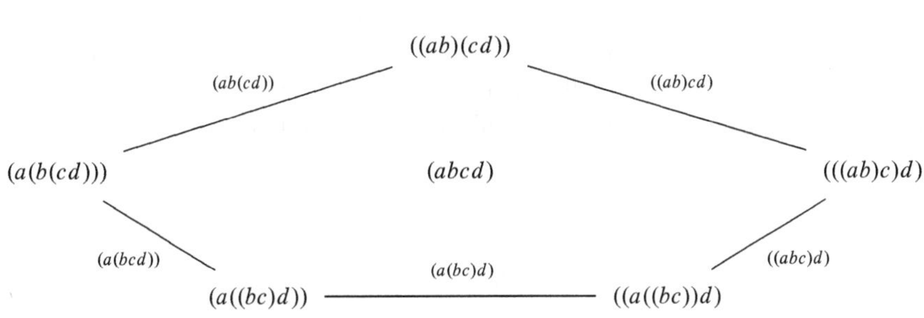

For each n ≥ 1, we denote K_n the associahedron for n elements. It is a cell complex of dimension n − 2. The cells of K_n are indexed by rooted planar trees with n leaves. The leaves are the elements in the string, in the same order. Each interior vertex of the tree (i.e., a vertex of order greater than 1) corresponds to a pair of matched parentheses around all the leaves that emanate from that vertex. The vertex just above the root then corresponds to the pair of parentheses around all the entire string. The cell associated with a planar rooted tree is the product of the cells associated with the constituent rooted trees with a single interior vertex. The top cell of the associahedron corresponds to the rooted tree with only one interior vertex. The boundary of the top cell is a union of closed codimension-1 cells; these correspond to rooted trees with two interior vertices. The cells associated to a planar tree T with s interior vertices of order $k_1 \ldots, k_s$ are identified with

$$K_{k_1-1} \times \cdots \times K_{k_s-1}$$

and hence give a product of cells whose dimension is $\sum_{i=1}^{s}(k_i-3)$. Since $\sum_{i=1}^{s} k_i = n+2s-1$, this product cell has codimension $s-1$. Also, the union of the interior edges of T, those that have neither a leaf nor the root as a vertex, forms a tree (the tree obtained by pruning the exterior edges that have a vertex of order 1 from the tree) so that T has $s-1$ interior edges.

There is a "dual" description which gives a subdivision of the cell structure K_n into a cubical complex. The cubes of this subdivision are indexed by the same collection of rooted planar trees. If one of the trees, say T, indexes a cell K(T) of K_n of dimension $n-2-r$, then T has r interior edges. In the cubical decomposition the tree T indexes a cube C(T) of dimension r. The cube C(T) is viewed as $[0,\infty]^{I(T)}$, where I(T) is the set of interior edges of T. Let T′ be another of the trees indexing a cell K(T′) of K_n. Then K(T) is a face of K(T′) if and only if T′ is obtained from T by collapsing some of its interior edges. In this case we identify C(T′) with the cubical face of C(T) defined by setting the cubical coordinates indexed by the edges of T that are collapsed in passing to T′ to zero. These inclusions generate an equivalence relation on the disjoint union of the C(T) as T ranges over the trees indexing the cells of K_n, and the quotient space is a cubical complex which can be identified with K_n in a natural way so that the cubical structure subdivides the cell structure of K_n. Notice that all the cubes meet at the vertex where all coordinates are zero: this is the barycenter of K_n. The boundary of K_n is the union of the faces of the top dimensional cubes, $C(T_{max})$, where at least one of the coordinates is ∞. The open cell in the boundary of K_n that contains any such point is the one associated to the tree obtained by collapsing all the interior edges of T_{min} whose coordinate is finite. Below is the picture for the resulting cubical decomposition of K_4.

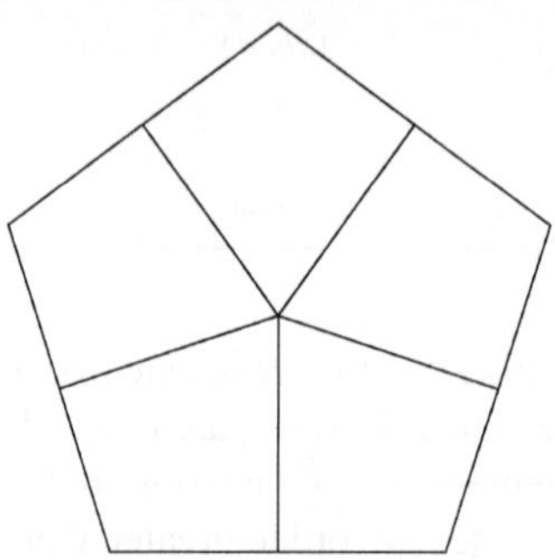

Cubical decomposition of K_4

The collection of associahedra $\mathcal{K} = \{K_n\}_n$ forms an operad of topological spaces (indeed of cell complexes). The maps $\mathcal{O}[(n, j); i]$ are given by the natural inclusion of $K_n \times K_j$ as the codimension-1 face of K_{n+j} indexed by the rooted tree obtained from the rooted tree with j leaves and one interior vertex and the rooted tree with n leaves and one interior vertex by attaching the root of the former to the ith leaf of the latter. This operad is compatible with the cubical decompositions given in the previous paragraph.

A space X is an *algebra over the associahedron operad* if there are continuous maps $K_n \times X^n \to X$ compatible with the structure maps of the operad. Said another way, *End*(X) is naturally an operad of topological spaces and X is an algebra over the associahedron operad if there is a map of topological operads $\mathcal{K} \to End(X)$. These objects are also called $\mathcal{A}_\infty$*-spaces.*

Stasheff not only introduced the associahedra, he proved that an $\mathcal{A}_\infty$-space X has a classifying space BX, so that an $\mathcal{A}_\infty$-space is a loop space (up to homotopy equivalence). In fact, the $\mathcal{A}_\infty$-structure on X determines BX and the homotopy equivalence $\Omega BX \to X$ up to homotopy; see pp. 48–58 in [24]. The converse is easy to establish, so this gives a characterization of which spaces are loop spaces; they are spaces that have $\mathcal{A}_\infty$-structures.

18.2 $\mathcal{A}_\infty$-Algebras and $\mathcal{A}_\infty$-Categories

There is an algebraic version of $\mathcal{A}_\infty$-spaces called $\mathcal{A}_\infty$-algebras. These algebras can be defined as algebras (in the category of graded vector spaces) over the operad in the induced category of differential graded vector spaces consisting of cellular chain complexes of the associahedra with the induced structure maps from the associahedra operad (thought of as an operad in the category of cell complexes and cellular maps). It turns out that there is a more elementary way to say all of this.

Definition. An $\mathcal{A}_\infty$-algebra is a graded vector space V, and for every $n \geq 1$, a map

$$m_n : V^{\otimes n} \to V$$

of degree $2 - n$ such that the following hold when we denote m_1 (which is of degree $+1$) by d:

$$(-1)^n dm_n =$$
$$\sum_{s=1}^{n} (-1)^{|a_1|+\cdots+|a_{s-1}|} m_n(a_1, \ldots, a_{s-1}, d(a_s), a_{s+1}, \ldots, a_n) +$$
$$\sum_{\substack{i+j=n+1 \\ i,j\geq 2}} \sum_{s=0}^{n-j} (-1)^p m_i(a_1, \ldots, a_{s-1}, m_j(a_s, \ldots, a_{s+j}), a_{s+j+1}, \ldots, a_n)$$

where the sign is given by $p = j + s(j+1) + j(|a_1| + \cdots + |a_{s-1}|$. Thus, m_n is associated with the rooted tree with n-leaves and only one interior vertex, and we see that the commutator $[m_n, d]$ is given by summing over the codimension-1 cells of the boundary of this associahedron of the operations associated with the corresponding rooted trees, those with exactly two interior vertices. It follows that the operation $d = m_1$ satisfies $d^2 = 0$, justifying the notation.

There is a corresponding categorical notion in any graded, tensor category: replace V by an object of the category and replace the m_n by morphisms $V^{\otimes n} \to V$ of degree $2-n$.

Definition. A *morphism* of A_∞-algebras $F\colon (M, m_i) \to (M', m'_i)$ is a sequence of homomorphisms

$$f_i\colon M^{\otimes i} \to M'$$

for $i \geq 1$ of degree $1-i$ satisfying

$$\sum_{k=0}^{i-1}\sum_{j=1}^{i-k} \pm f_{i-j+1}(a_1 \otimes \cdots \otimes a_k \otimes m_j(a_{k+1} \otimes \cdots \otimes a_{k+j}) \otimes \cdots \otimes a_i) =$$

$$\sum_{s=1}^{i} \sum_{\sum_1^t k_r = i} m'_s(f_{k_1}(a_1 \otimes \cdots \otimes a_{k_1}), \ldots, f_{k_s}(a_{i-k_s+1} \otimes \cdots \otimes a_i))$$

These objects were first introduced in the 1960s by May and Stasheff (see [16] for some a history of these ideas), among others. A_∞-structures have seen a resurgence recently with applications to low-dimensional topology, to symplectic topology, and quantum field theory and string theory.

There is a variant of the notion of an A_∞-algebra which is an A_∞-category. For an A_∞-category has a collection of objects and for objects A and B, there is a graded vector space of morphisms Hom(A, B). For each $n \geq 1$, there is a morphism

$$m_n\colon \mathrm{Hom}(A_1, A_2) \otimes \mathrm{Hom}(A_2, A_3) \otimes \cdots \otimes \mathrm{Hom}(A_n, A_{n+1}) \to \mathrm{Hom}(A_1, A_{n+1})$$

of degree $2-n$. This satisfy the same algebraic axioms as for an A_∞ algebra. In particular, $d = m_1$ is a degree $+1$ map of square zero (i.e., a differential) on Hom(A, B).

The motivating example occurs in [6]. The objects formalized from what they did are called *Fukaya categories*. The motivating example was to formalize the algebraic structure associated with the Lagrangian Floer homology of a symplectic manifold. This class of examples has been much studied for the insight it gives into symplectic topology; it is also crucial in the refined formulations of homological mirror symmetry introduced by Kontsevich [13].

As we shall see in considering Lagrangian Floer homology, it is natural to replace rooted planar trees with n leaves by a disk with $n+1$ marked points on its boundary, one of which is distinguished as the root, or the output, and the other n are the inputs, ordered counterclockwise from the root. Given a symplectic manifold M, we form an A_∞-category whose objects are Lagrangian submanifolds of M. For Lagrangian submanifolds L and L′ by definition, Hom(L, L′) is the Floer cochain complex CF(L, L′). The generators of this chain complex are the points of intersection $L \cap L'$ (which we assume to be transverse). We also assume that the cochain complex has an integral grading. Then the co-boundary map, or differential, is defined as follows: Given intersection points a and b with the degree of b being one more than the degree of a, consider all holomorphic maps to the disk minus 2 boundary points into M with one boundary arc in L and the other boundary arc in L′, with the maps converging at the missing points to a and b. Modulo the action of $\mathbb{R}$, there will be a finite number of these. Their algebraic count is the coefficient of b in d(a). The higher maps m_n are defined as follows. Given n + 1 Lagrangian submanifolds $L_1, \ldots, L_{n+1}$ with all intersections being transversal, we define the higher map

$$m_n: CF(L_1, L_2) \otimes \cdots \otimes CF(L_n, L_{n+1}) \to CF(L_1, L_{n+1})$$

by counting finite energy holomorphic maps from the disk minus n + 1 boundary points into M with the boundary arcs, in order, mapping to $L_1, \ldots, L_{n+1}$. Under suitable hypotheses (see [20] for one such context), these structure maps produce an A_∞-category structure, which is the Fukaya category of this symplectic manifold.

18.3 C_∞-Algebras and DGAs

After this excursion into homotopy theory and symplectic geometry, let us turn to the relationship of these structures to rational homotopy theory. To do this we introduce the graded-commutative analogue of A_∞-algebras; see [10]. These are the so-called C_∞-algebras. They are the quotient of A_∞-algebras when one forces graded commutativity. There is a shuffle product on the tensor algebra, $T^*(V)$, of a graded vector space V. This product is denoted

$$\mu_{sh}: V^{\otimes m} \otimes V^{\otimes n} \to V^{\otimes (m+n)}$$

and it is given by the formula

$$\mu_{sh}\left((a_1 \otimes \cdots \otimes a_m) \otimes (a_{m+1} \otimes \cdots \otimes a_{n+m})\right) = \sum \pm a_{\sigma(1)} \otimes \otimes a_{\sigma(n+m)},$$

where the sum is over all (m, n)-*shuffles*, i.e., all permutations σ of $\{1, \ldots, m+n\}$ with the property that if $1 \leq \sigma(i) < \sigma(j) \leq m$ or if $m+1 \leq \sigma(i) < \sigma(j) \leq m+n$, then $i < j$. This exactly forces both the set $a_1, \ldots, a_m$ and the set $a_{m+1}, \ldots, a_{m+n}$ to be in increasing order after the shuffle. The sign associated with σ is the product of the sign of σ as a permutation and the Kozul sign of the associated with the re-ordering of the a_i. (The Kozul sign for a switch of two neighboring elements $a \otimes b \mapsto b \otimes a$ is $(-1)^{|a|\cdot|b|}$; this generates the Kozul sign for arbitrary permutations.)

Definition. A C_∞-*algebra* is an A_∞ algebra (M, m_i) with the property that each of the operations m_i, $i \geq 2$, annihilates the image of all shuffles:

$$m_i \left(\mu_{sh}\left((a_i \otimes \cdot a_j) \otimes (a_{j+1} \otimes \cdots a_i)\right)\right) = 0$$

for all $1 \leq j < i$. A morphism $F = \{f_i\}$ of C_∞-algebras is a morphism of the A_∞-algebras with the property that every f_i vanishes on the image of all shuffles.

C_∞-algebras can be used to give another algebraic description of rational homotopy theory. Let A^* be a DGA. We can view this as a C_∞-algebra with $m_1 = d$, with m_2 being the multiplication in the algebra, and with $m_i = 0$ for all $i > 2$. We have the following theorem of Kadeishvili [10]. It says that up to homotopy equivalence, we can collapse A^* onto its cohomology at the expense of endowing the cohomology with a possibly nontrivial C_∞-structure.

Theorem 18.1. *Let* A^* *be a DGA defined over* $\mathbb{Q}$. *Then, there is a* C_∞*-algebra* $(H^*(A), \bar{m}_i))$ *with* $m_1 = 0$ *and with* $\bar{m}_2$ *being the multiplication induced on* $H^*(A)$ *from the multiplication in* A^* *and for which there is a morphism of* C_∞*-algebras from* $(H^*(A), \bar{m}_i)) \to A^*$ *inducing the identify on cohomology. Furthermore, this* C_∞*-algebra structure is unique up to isomorphism in the category of* C_∞*-algebras. Lastly, two DGAs have isomorphic minimal models if and only if the associated* C_∞*-structures on their cohomology groups are isomorphic in the category of* C_∞*-algebras. In particular, a DGA* A^* *is formal if and only if the* C_∞*-structure on its cohomology* $(H^*(A(, \bar{m}_i))$ *is isomorphic to a* C_∞*-algebra with* $\bar{m}_i = 0$ *for* $i \neq 2$.

We shall not prove this theorem, but as a hint as to the techniques used, we show how to compute Massey products from a C_∞-algebra on cohomology. First let us examine the maps $f_i \colon H^*(A)^{\otimes i} \to A^*$ (of degree $1 - n$) which form the morphism of C_∞-algebras. The map f_1 satisfies $df_1 = 0$. This implies that for each cohomology class a, $f_1(a)$ is a closed element in A^*. The statement that the morphism induces the identity on cohomology means that $f_1(a)$ is a closed element representing the cohomology class a. Now consider $f_2(a \otimes b)$. We have $df_2(a \otimes b) = m_2(f_1(a) \otimes f_1(b)) - f_1(\bar{m}_2(a \otimes b))$. That is to say, $d(f_2(a \otimes b))$ is the difference between the product of the representative closed elements for a and b and the representative closed element for $a \cup b$. In the special case when $a \cup b = 0$, we see that $d(f_2(a \otimes b)$ is the cup product of the representatives closed elements for the classes a and b. Lastly, again using the fact that $m_3 = 0$, we have

$$\begin{aligned} 0 &= f_1(\bar{m}_3(a \otimes b \otimes c)) \\ &= m_2(f_2(a \otimes b)) \otimes f_1(c)) \pm m_2(f_1(a) \otimes f_2(b \otimes c)) \pm d(f_3(a \otimes b \otimes c)). \end{aligned}$$

In the special case when $a \cup b = b \cup c = 0$, it follows that $\bar{m}_3(a \otimes b \otimes c)$ is a cohomology class in $H^{|a|+|b|+|c|-1}(A^*)$ whose equivalence class modulo the indeterminacy is the Massey triple product $< a, b, c >$. These ideas generalize to allow one to construct a C_∞-isomorphism from a C_∞-structure on $H^*(A^*)$ to A^*, an isomorphism inducing the identity on cohomology.

There is an analogous theorem in the noncommutative case that says a (not necessarily commutative) DGA is equivalent in an appropriate homotopy category to an A_∞-algebra structure on its cohomology.

Chapter 19
Exercises

1. If X, Y are finite CW complexes, then prove that $X \times Y$ is also. Show that the k-skeleton

$$(X \times Y)^{(k)} = \bigcup_{i+j=k} X^{(i)} \times Y^{(j)}$$

From this, deduce that the cellular chain groups

$$\tilde{C}_*(X \times Y) \cong \tilde{C}_*(X) \otimes \tilde{C}_*(Y)$$

with boundary

$$\partial_{X \times Y} = \partial_X \otimes 1 \pm 1 \otimes \partial_Y.$$

Using this, prove the Künneth theorem.
Why is the Künneth theorem hard for singular or simplicial homology?
2. Let (X, A) be a CW pair where A is a finite subcomplex. Show that

$$X/A = \{X \text{ with A collapsed to the base point}\}$$

is again a CW complex. Is the same thing true if A is infinite? Show that

$$H_*(X, A) \cong \tilde{H}_* (X/A) \text{ (use excision)}$$
$$\pi_n(X, A) \neq \pi_n(X, A) \text{ in general}$$

Hint: Take X = 2-disk and A = S^1 its boundary. Then, $X/A \cong S^2$ and so $\pi_3(X/A) \neq 0$. What about $\pi_3(X, A)$?)
3. (a) Show that $\pi_0(X) = \{\text{set of path components of } X\}$.
   (b) Let ΩX be the loop space of X. Show that composition of loops induces a group structure on $\pi_0(\Omega X)$. Show that

P. Griffiths and J. Morgan, *Rational Homotopy Theory and Differential Forms*, Progress in Mathematics 16, DOI 10.1007/978-1-4614-8468-4_19,

© Springer Science+Business Media New York 2013

$$\pi_0(\Omega X) \cong \pi_1(X).$$

(c) More generally, prove that

$$\pi_{n-1}(\Omega X) \cong \pi_n(X),$$

and deduce that if by induction we define $\Omega^n X = \Omega(\Omega^{n+1} X)$, then

$$\pi_0(\Omega^n X) \cong \pi_n(X).$$

4. Let X be a CW complex and $f : S^n \to X^{(n)}$ a map. Let $Y = X \cup_f e^{n+1}$ be the CW complex obtained by attaching $(n+1)$-cell by f. Show that the homotopy type of Y depends only on the homotopy class

$$[f] \in \pi_n(X^{(n)}).$$

Note: We are not considering [f] in $\pi_n(Y)$. Why not?

5. Given spaces X,Y, define the *wedge*

$$X \vee Y = (X \times y_0) \cup (x_0 \times Y) \subset X \times Y.$$

Compute the cohomology *ring* of $\mathbb{C}P^{n-1} \vee S^{2n}$. Is this the same as the cohomology ring of $\mathbb{C}P^n$?

6. Using the preceding exercise, show that

$$\pi_{2n-1}(\mathbb{C}P^{n-1}) \neq 0.$$

7. Using the exact homotopy sequence of a fibration, show that

$$\pi_n(X \times Y) \cong \pi_n(X) \oplus \pi_n(Y).$$

8. Let X be a space, $\tilde{X} \xrightarrow{f} X$ its universal covering. Using the homotopy lifting property, prove that

$$\pi_n(\tilde{X}) \xrightarrow{f_*} \pi_n(X)$$

is an isomorphism for $n \geq 2$. Using this, let

X = 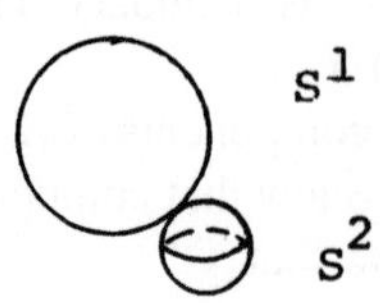

(= 2 sphere with wire attached).

Show that $\pi_2(X) \cong \mathbb{Z} \oplus \mathbb{Z} \oplus \dots$ infinite number of times. Thus, the homotopy groups of a finite complex may be infinitely generated. [Is π_1 of a finite complex finitely generated?]

9. Consider the space $X = \{x = 0\} \cup \{y = \sin \frac{1}{x}\}$ in the (x, y)-plane. Show that $\pi_0(X)$ has two elements, but X is connected as a topological space. Let $Y = \{a\} \cup \{b\}$ be a space with two distinct points and $f : Y \to X$ the map

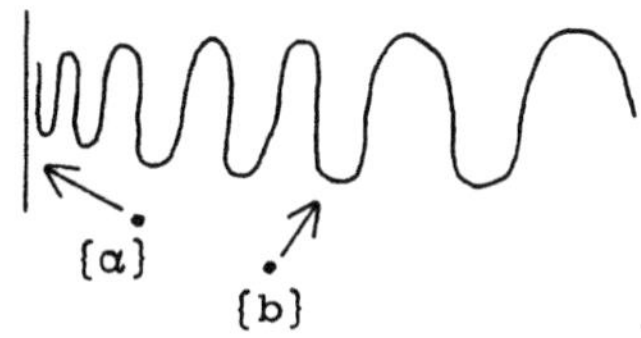

Show that f_* is an isomorphism on homotopy groups, but f^{-1} does not exist in the homotopy category. Thus, the Whitehead theorem is false for non-CW complexes.

10. (a) Assuming that $n > 1$, show that the inclusion

$$S^n \vee S^n \hookrightarrow S^n \times S^n$$

induces an isomorphism

$$\pi_n(S^n \vee S^n) \cong \pi_n(S^n \times S^n).$$

(Hint: Using Exercise 7, show that the map is onto. If we have

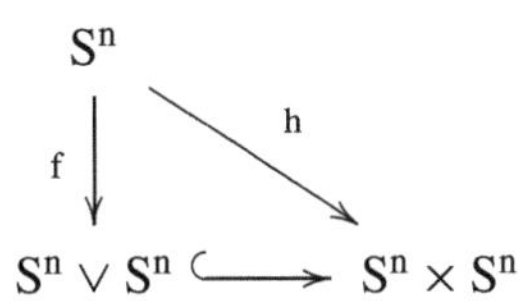

such that h is homotopic to a constant, then since $\dim(S^n \times S^n) = 2n > n + 1$, we may assume that the homotopy misses a point $p \in S^n \times S^n$ (why? cf. the simplicial approximation theorem). Thus, you must prove that

Lemma. $S^n \times S^n \setminus \{p\}$ *retracts onto* $S^n \vee S^n$.

For this, write $S^n = I^n/\partial I^n$ and use a picture like

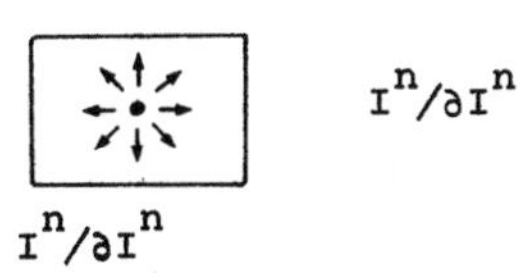

(b) What is $\pi_1(S^1 \vee S^1)$ (picture = ∞)?

11. (a) Fill in the details of the proof of Theorem 3.2.
    (b) Why does the proof fail for n = 1?

12. Given $A = B \cup_f e^{n+1}$, show that

$$\pi_i(B) \cong \pi_i(A) \qquad i < n$$

$$\pi_n(B) \longrightarrow \pi_n(A) \longrightarrow 0$$

(Hint: Show that it is sufficient to check that any map $X^{(n)} \xrightarrow{f} A$ from an n-complex to A is homotopic to a map missing a point in e^{n+1}. Now why is this true? See Exercise 10 above.)

13. Using Exercise 4 and the simplicial approximation theorem, show that any (finite) CW complex has the same homotopy type as a simplicial complex. (Hint: Given X, suppose that we have an n-dimensional simplicial complex $Y^{(n)}$ and a homotopy equivalence $X^{(n)} \underset{g}{\overset{\sim}{\rightarrow}} Y^{(n)}$. Let $X^n \cup_f e^{n+1}$ be an n+1 cell attached to X^n. Make a simplicial approximation h to $g \circ f$ and check that $Y^{(n)} \cup_h e^{n+1}$ is a simplicial complex and there is a homotopy equivalence

$$X \cup_f e^{n+1} \sim Y^{(n)} \cup_h e^{n+1}.$$

Now continue adding more cells.)

14. A space X is of *category two* if

$$X = X_+ \cup X_- (X_+,\ X_- \text{ open in } X)$$
$$X_+, X_- \text{ are contractible to a point.}$$

Show that, for any space Y, the *suspension* SY is of category two. In particular, any sphere S^n is of category two (this is obvious).

15. Let X be of category two and $Y = X_+ \cap X_-$.
    Show that the sets

$$\mathrm{Vect}^r(X)$$

and

$$[Y, GL(r, \mathbb{C})]$$

are in 1–1 correspondence (here $\mathrm{Vect}^r(X)$ is the set of isomorphism classes of vector bundles over X with fiber $\mathbb{C}^r$).
Hint: Setting $E_+ = E|X_+$ and $E_- = E|X_-$, both $E_+ \to X_+$ and $E_- \to X_-$ are trivial (why?). Now compare the trivializations over $X_+ \cap X_- = Y$.)

16. An r-*frame* in $\mathbb{C}^N$ is given by r orthonormal vectors $(e_1, \ldots, e_r)$. The set of r-frames in $\mathbb{C}^N$ is the *Stiefel manifold* $S(r, N)$. Show that

$$\pi_i(S(r,N)) = 0 \qquad i < 2N - 2r + 1$$
$$\pi_{2N-2r+1}\,(S(r,N)) \cong \mathbb{Z}.$$

Hint: Observe first that $S(1, N) = S^{2N-1}$ is the unit sphere $\mathbb{C}^N$, so the result is correct for $r = 1$. Now use induction on r and the exact homotopy sequence of the fibering

$$\begin{array}{ccc} S^{2N-2r+1} & \longrightarrow & S(r,N) \\ & & \downarrow \pi \\ & & S(r-1,N) \end{array}$$

where π is given by $\pi(e_1, \ldots, e_r) = (e_1, \ldots, e_{r-1})$.

17. Show that the set of all r-frames in $\mathbb{C}^r$ is the unitary group $U(r)$ (the r-frames in $\mathbb{C}^r$ are just the orthonormal bases).

Deduce the fibration

$$\begin{array}{ccc} U(r) & \longrightarrow & S(r,N) \\ & & \downarrow \pi \\ & & G(r,N) \end{array}$$

here $G(r, N)$ is the Grassmann manifold of r-planes in $\mathbb{C}^N$ and

$$\begin{aligned} \pi\,(e_1, \ldots, e_r) &= e_1 \wedge \ldots \wedge e_r \\ &= r - \text{plane spanned by } e_1, \ldots, e_r. \end{aligned}$$

Using this fibration, show that

$$\pi_i(G(r,N)) \cong \pi_{i-1}(U(r)) \quad \text{for } i < 2N - 2r + 1.$$

18. Use the inclusion $U(n) \hookrightarrow U(n+1)$ given in matrices by

$$A \longrightarrow \begin{pmatrix} A & 0 \\ 0 & 1 \end{pmatrix}$$
$$A = n \times n \text{ unitary matrix.}$$

Deduce that there is a fibering

$$\begin{array}{ccc} U(n) & \longrightarrow & U(n+1) \\ & & \downarrow \pi \\ & & S^{2n+1} \end{array}$$

and conclude that

$$\pi_i(U(n)) \cong \pi_i(U(n+1)) \quad \text{for} \quad i < 2n.$$

Thus, the homotopy of the unitary group is *stable* in the sense that

$$\pi_i(U(n)) \cong \pi_i(U(n+m)) \quad (i < n \text{ and } m \geq 0).$$

Letting $U = \lim_{n\to\infty} U(n)$, it is a famous theorem of Bott (*periodicity theorem*) that

$$\begin{cases} \pi_{2i+1}(U) \cong \mathbb{Z} \\ \pi_{2i}(U) = 0 \end{cases}$$

In Chap. 13 this theorem is proved in the rational homotopy category.

Note: Let $G(n, 2n)$ be the Grassmannian of n-planes in $\mathbb{C}^{2n}$. Using Exercise 17 and Bott periodicity, we have

$$\pi_2 i(G(n, 2n)) \cong \mathbb{Z} \qquad (n >> i)$$
$$\pi_{2i-1}(G(n, 2n)) = 0$$

The map $\pi_{2i}(G(n, 2n)) \xrightarrow{\rho} \mathbb{Z}$ is given as follows: A homotopy element in $\pi_{2i}(G(n, 2n))$ is given by $S^{2i} \to G(n, 2n)$. Pulling back the universal bundle ζ on $G(n, 2n)$ gives a vector bundle $\mathbb{C}^n \to E \to S^{2i}$. Then,

$$\rho(f) = c_i(E)/(i-1)!$$

is an element of integral homology where $c_i(E\,) \in H^{2i}(S^{2i})$ is the ith *Chern class* of E, and we have identified $H^{2i}(S^{2i}) \cong \mathbb{Z}$. This divisibility property of the Chern classes of a vector bundle is extremely subtle and is the sort of thing <u>not</u> covered by rational homotopy theory.

19. Complete the proof of Corollary 4.6 as follows: Given a CW complex X with $\pi_1(X) = 0$ and $H_i(X) = 0$ for $i > n$ and $H_n(X)$ free abelian, look at the $(n-1)$-skeleton $X^{(n-1)}$. Using that,

$$H_i(X^{(n-1)}) \cong H_i(X) \quad i \leq n-2$$
$$H_{n-1}(X^{(n-1)}) \longrightarrow H_{n-1}(X) \longrightarrow 0$$

deduce the exact sequence

$$0 \longrightarrow H_n(X) \longrightarrow H_n(X,\ X^{(n-1)}) \longrightarrow H_{n-1}(X^{(n-1)}) \xrightarrow{i} H_{n-1}(X) \longrightarrow 0.$$

Now $H_{n-1}(X^{(n-1)})$ is free abelian. (why?) and so is $H_n(X)$ by assumption. Thus, $H_n(X, X^{(n-1)})$ is free abelian (why?). Show finally that $\pi_i(X, X^{(n-1)}) = 0$ for $i \leq$ n-1 and $\pi_n(X, X^{(n-1)}) \cong H_n(X, X^{(n-1)})$, so that we may attach n-cells $\{e^n_\alpha\}$ to have $Y = X^{(n-1)} \cup_{f_\alpha} \{e^n_\alpha\}$ together with a map $Y \to X$ inducing an isomorphism on homology.

20. Here is an example of CW complexes X, Y and a map

$$X \xrightarrow{f} Y$$

such that f_* is isomorphism on $H_*(X) \xrightarrow{\cong} H_*(Y)$ but f is not a homotopy equivalence. Consider

$Y_1 = S^1 \vee S^2$ 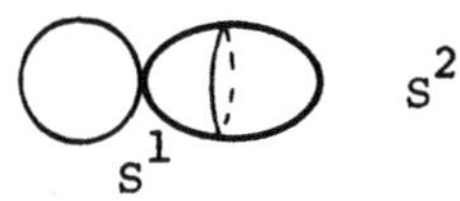

The universal covering $\tilde{Y}_1$ has a picture

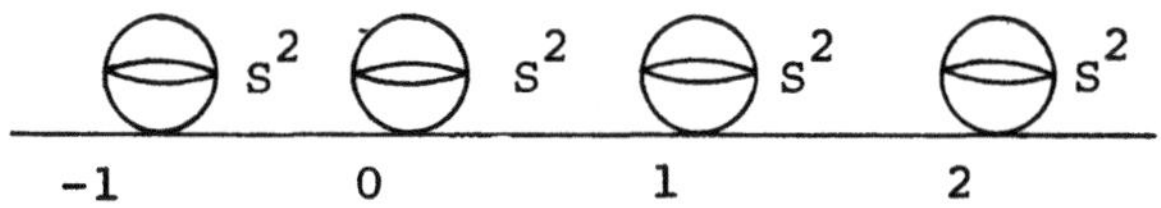

where $\mathbb{R}$ is the real line (= universal covering of S^1) and 2-spheres are attached at each integer point. Let $T : \tilde{Y}_1 \to \tilde{Y}_1$ be translation by 1 on $\mathbb{R}$ viewed as an automorphism of $\tilde{Y}_1$. Thus, $Y_1 = \tilde{Y}_1/\{T\}$ where T generates $\pi_1(Y_1)$ as a group of covering transformations on $\tilde{Y}_1$.
Attach a 3-cell to $\tilde{Y}_1$ by a map

$$S^2 \xrightarrow{\tilde{f}} \tilde{Y}_1$$

and let the attaching map to Y_1 be $\pi \circ \tilde{f} = f$

$$\begin{array}{ccc} S^2 & \xrightarrow{\tilde{f}} & \tilde{Y}_1 \\ & \searrow^{f} & \downarrow \pi \\ & & Y_1 \end{array}$$

To give $\tilde{f}$, we map S^2 to $S^2_{\{T=0\}}$ with degree +2 and to $s^2_{\{T=1\}}$ with degree −1

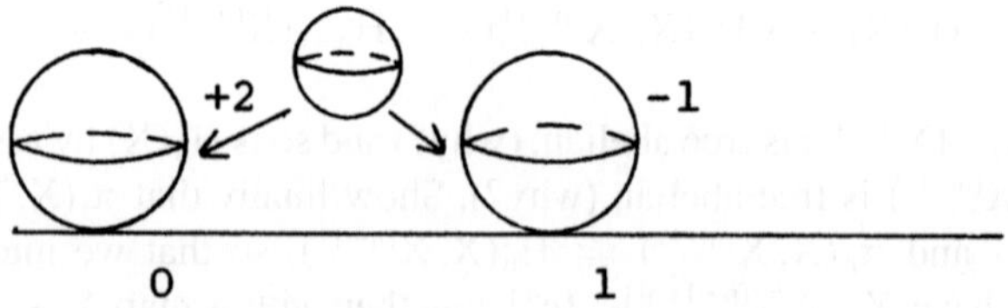

Let $Y = Y_1 \cup_f e^3$. Then,

$$H_i(Y) = 0 \qquad i \neq 1$$

$$H_1(Y) \cong \mathbb{Z} \quad \text{(why?)},$$

and so the inclusion map

$$X = S^1 \overset{i}{\hookrightarrow} Y$$

induces an isomorphism on H_* and on π_1, but $\pi_2(X) = 0$, and we will check that $\pi_2(Y) \neq 0$ so that i is not a homotopy equivalence. Observe that the homotopy groups $\pi_i(\tilde{Y}_1)$ are $\pi_1(Y_1)$-modules (why?) (the homology $H_*(\widetilde{Y_1})$ is also a $\pi_1(Y_1)$ module). Letting g be the generator of $\pi_2(\tilde{Y}_1)$ corresponding to the S^2 at 0, we have as sets

$$\pi_2\left(\tilde{Y}_1\right) = \{T^n g^m\}_{n,m\in\mathbb{Z}} \quad \text{(why?)}.$$

Thus, as groups

$$\pi_2\left(\tilde{Y}_1\right) \cong \mathbb{Z}[T] \quad \text{(why?)}.$$

Now

$$\pi_2(Y) \cong \pi_2(\tilde{Y}) \cong H_2(\tilde{Y}) \quad \text{(why?)}.$$

Finally,

$$H_2(\tilde{Y}) \cong \mathbb{Z}[T]\,/2T - 1 \neq 0$$

(since $\partial e^3 = 2S_0^2 - S_1^2$) (why?).

Remark. (i) There is a non-simply connected version of the Hurewicz theorem. It says that a map between two CW complexes, $f : X \to Y$, is a homotopy equivalence if and only if:

(1) $f_* : \pi_1(X) \to \pi_1(Y)$ is an isomorphism.

(2) $f_* : H_*(X; \mathbb{Z}[\pi_1(X)]) \to H_*(Y; \mathbb{Z}[\pi_1(Y)])$ is an isomorphism.

The point is that $H_*(X; \mathbb{Z}[\pi_1(X)])$ is identified with the ordinary integral homology of the universal cover $\tilde{X}$ of X. Thus, condition (2) implies that $\tilde{f} : \tilde{X} \to \tilde{Y}$ induces an isomorphism in homology and hence on homotopy groups (since $\pi_1(\tilde{X}) = \pi_1(\tilde{Y}) = 0$).

(ii) The dodecahedron group G is a *perfect group* (i. e., $G = [G, G]$) acting freely on S^3, and so $X = S^3/G$ is a 3-manifold with $H_i(X) = 0$ ($i \neq 0, 3$), $H_0(X) \cong H_3(X) \cong \mathbb{Z}$. Thus, X is a homology sphere, and the map

$$X \xrightarrow{f} S^3$$

obtained by collapsing the 2-skeleton of X to a point induces an isomorphism on homology. This example, due to Poincaré, gives another example showing that the homology does not suffice to determine the homotopy type of even a 3-manifold.

21. **Problem on algebraic Euler characteristics.** Recall that a finitely generated abelian group G has a *rank* $\rho(G) = \dim_{\mathbb{R}}(G \otimes_{\mathbb{Z}} \mathbb{R})$. Suppose that (C_n, ∂) is a chain complex with C_n finitely generated and $C_n = 0$ for $n < 0, n \geq n_0$. Set

$$c_n = \rho(C_n) = \text{ rank of } C_n$$
$$H_n = n^{\text{th}} \text{ homology of } \{C_*, \partial\}$$
$$b_n = \rho(H_n) = n^{\text{th}} \text{ Betti number.}$$

Prove the relation (*Eule–Poincaré-Hopf*)

$$\Sigma_n(-1)^n c_n = \Sigma_n (-1)^n b_n.$$

Hint: Use induction on n_0 and the fact that if

$$0 \to G' \to G \to G'' \to 0$$

is an exact sequence of finitely generated abelian groups, then

$$\rho(G) = \rho(G') + \rho(G'').)$$

If X is a finite CW complex and

$$c_n = \rho(C_n(X)) = \# \text{ of n-cells in } X$$
$$b_n = \rho(H_n(X)) = n^{\text{th}} \text{ Betti number of } X$$
$$\chi(X) = \Sigma_n(-1)^n b_n = \underline{\text{Euler characteristic of}} \ X,$$

then you have proved

$$\Sigma_n(-1)^n c_n = \chi(X) \quad \text{(formula of Euler–Poincaré).}$$

Exercises on Euler characteristic and spectral sequences.

22. Let $F \to E \to B$ be a fibration where $\pi_1(B)$ acts trivially on $H^*(F)$. Suppose that $H^*(B)$ and $H^*(F)$ are finitely generated. Prove that

(i) $H^*(E)$ is finitely generated.
(ii) $\chi(E) = \chi(B)\chi(F)$ where "χ" means Euler characteristic.

(Hint: Use the Serre spectral sequence to see that $H^*(E)$ can be "no larger" than $H^*(B) \otimes H^*(F)$, and so $H^*(E)$ is finitely generated. To prove (ii) use Exercise (21) together with the relation

$$E_{r+1} = \text{homology of } \{E_r, d_r\}.$$

The formula $\chi(E) = \chi(B)\chi(F)$ means that the Euler characteristic is multiplicative for fiber spaces.

23. Let $F \to E \to S^n$ be a fibration over an n-sphere ($n \geq 2$). Deduce the exact sequence (*Wang sequence*)

$$\ldots \to H^p(E) \xrightarrow{i_*} H^p(F) \xrightarrow{\mathscr{D}} H^{p-n+1}(F) \longrightarrow H^{p+1}(E) \xrightarrow{i*} \ldots$$

where $i : F \to E$ is the inclusion and where

$$\mathscr{D}(u \cdot v) = \mathscr{D}(u) \cdot v \pm u\mathscr{D}(v).$$

(Hint. Since $H^q(S^n) = 0$ for $q \neq 0$, n the E_2 term of the spectral sequence looks like

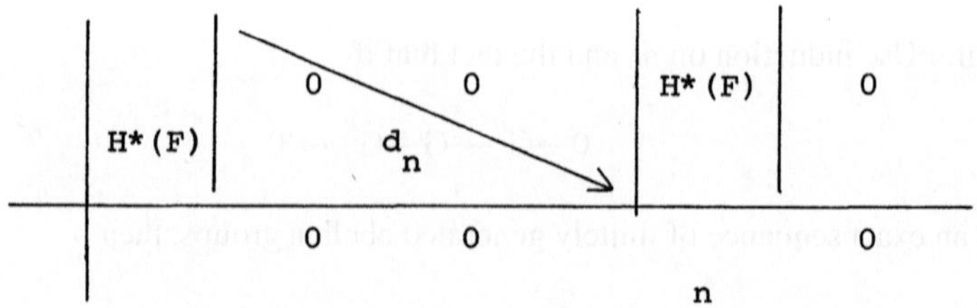

Deduce that the only nonzero differential in the spectral sequence is d_n. From this, conclude that

(a) $E_2 = \ldots = E_{n-1}, E_{n+1} = E_{n+2} = \ldots = E_\infty$
(b)

$$\begin{array}{ccccccccccc}
0 & \longrightarrow & H^p(E) & \xrightarrow{i^*} & H^p(F) & \longrightarrow & H^{p-n+1}(F) & \longrightarrow & H^{p+1}(E) & \longrightarrow & 0 \\
 & & \cup\uparrow & & =\uparrow & & =\uparrow & & \cup\uparrow & & \\
0 & \longrightarrow & E_\infty^{p,0} & \longrightarrow & E_n^{p,0} & \xrightarrow{d_n} & E_n^{p-n+1,n} & \longrightarrow & E_\infty^{p-n+1,n} & \longrightarrow & 0
\end{array}$$

The point of this exercise is to illustrate the general principle: When the E_2-term of the spectral sequence has most terms zero, then it simplifies considerably.)

24. **Another problem on spectral sequences.** Let $S^n \to E \xrightarrow{\pi} B$ be a fibration with fiber a sphere and where $\pi_0(B) = 0$ and $\pi_1(B)$ acts trivially on $H^*(S^n)$. Deduce the exact cohomology sequence (*Gysin sequence*)

$$\ldots \xrightarrow{\pi*} H^p(E) \longrightarrow H^{p-n}(B) \xrightarrow{\psi} H^{p+1}(B) \xrightarrow{\pi*} H^{p+1}(E) \to \ldots$$

where ψ is cup product with a class $e \in H^{n+1}(B)$ (e is the *Euler class*, defined by obstruction theory in Exercise (27) below).

(Hint. The E_2 term in the spectral sequence is

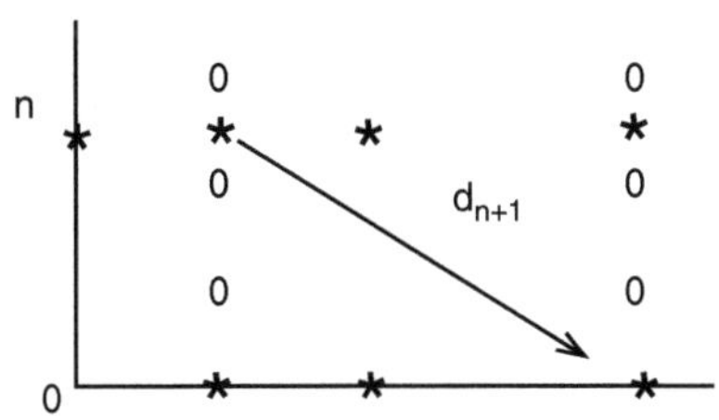

and so $E_2 = \ldots = E_n$, $E_{n+2} = \ldots = E_\infty$ and d_{n+1} is the only nonzero differential. The Euler class $e = d_{n+1}(1)$ where $1 \in H^n(F) \cong E_2^{0,n}$ is the generator of $H^n(S^n)$, and where now the argument proceeds as in the previous exercise using $d_n(\alpha \otimes \beta) = \alpha \otimes d_n(\beta)$ for $\alpha \in H^*(B)$, $\beta \in H^*(F)$).

25. **Thom isomorphism.** Let $E \to X$ be an orientable vector bundle of dimension n. Choose a metric on E. Let $D(E) \subset E$ be the unit disk sub-bundle with respect to this metric, and let S(E) be the unit sphere sub-bundle. We have a relative fibration $(D(E), S(E)) \to X$; the fibers are pairs (D^n, S^{n-1}). There is a relative version of the Serre spectral sequence where $E_2^{p,q} = H^p(X; H^q(D^n, S^{n-1}))$ and $E_\infty \Rightarrow H^*(D(E), S(E))$ Using this, establish the *Thom isomorphism*:

There is a class $U \in H^n(D(E), S(E))$ so that

$$H^*(X) \cong H^*(D(E)) \xrightarrow{\cup U} H^*(D(E), S(E))$$

is an isomorphism.

26. **Still another problem on spectral sequences.** Let $\mathbb{P}^n \to E \to B$ be a fibration with complex projective space as fiber.

Recall that

$$H^*(\mathbb{P}^n) \cong \mathbb{Z}[x]/(x^{n+1}), \quad x \in H^2(\mathbb{P}^n).$$

Suppose that there exists a class $\zeta \in H^2(E)$ such that $\zeta|\mathbb{P}^n = x$. (Here $\zeta|\mathbb{P}^n$ means restriction of ζ to fiber). Show that we have the ring isomorphism

$$H^*(E) \cong H^*(B)[\zeta]/(\zeta^{n+1} - \Sigma_{q=1}^{n+1}(-1)^{n+1-q}c_q\zeta^{n+1-q})$$

where $c_q \in H^{2q}(B)$.

(Hint: There is a natural map

$$H^*(B)[\zeta] \longrightarrow H^*(E)$$

coming from $H^*(B) \xrightarrow{\pi^*} H^*(E)$ and the ring structure on $H^*(E)$. By assumption there exists $\zeta_0 \in E_\infty^{0,2}$ corresponding to $x \in E_2^{0,2} \cong H^2(\mathbb{P}^n)$. Thus,

$$d_2X = d_3X = \ldots = 0.$$

Since $E_2 \cong H^*(B)[x]/(x^{n+1})$, deduce that $E_2 = E_3 = E_4 = \ldots = E_\infty$. Thus

$$H^*(B)[\zeta\,] \longrightarrow H^*(E)$$

is surjective. Additively there is an isomorphism

$$H^*(E) \cong \oplus_{q=0}^n H^*(B)\zeta^q,$$

and so the kernel of the restriction mapping

$$H^*(E) \longrightarrow H^*(P^n)$$

is $\oplus_{q=1}^n H^*(B)\zeta^q$. But $\zeta^{n+1}|\mathbb{P}^n = 0$, and so we obtain a relation in the ring $H^*(E)$ of the form

$$\zeta^{n+1} = \Sigma_{q=1}^{n+1}(-1)^{n-q}c_q\zeta^{n+1-q}, \quad c_q \in H^{2q}(B).$$

Deduce finally that

$$H^*(B)[\zeta]/(\zeta^{n+1} - \Sigma_{q=1}^{n+1}(-1)^{n+1-q} c_q \zeta^{n+1-q}) \longrightarrow H^*(E)$$

is an injection.)
(Note: The classes $c_q \in H^{2q}(B, \mathbb{Z})$ are Chern classes as explained in Exercise (30) below. This definition, due to Grothendieck, is the easiest.)

27. **A final problem on spectral sequences.** Let U_n be the unitary group of $n \times n$ matrices A satisfying $A^{tr}A = I$. Prove that the cohomology ring

$$H^*(U_n) \cong \wedge(x_1, x_3, \ldots, x_{2n-1})$$

is an exterior algebra with generators in dimensions 1, 3, ..., $2n - 1$. Thus, there is an isomorphism of rings

$$H^*(U_n) \cong H^*(S^1 \times S^3 \times \ldots \times S^{2n-1}).$$

Hint: Writing $A = (e_1, \ldots, e_n)$ where the e_i are column vectors in $\mathbb{C}^n$, the unitary condition is that $e_1, \ldots, e_n$ should give a unitary frame.
The map

$$U_n \longrightarrow S^{2n-1}$$
$$A \mapsto e_1$$

gives a fibration

$$\begin{array}{ccc} U_{n-1} & \longrightarrow & U_n \\ & & \downarrow \\ & & S^{2n-1} \end{array}$$

(why?). By induction we have $H^*(U_{n-1}) \cong \wedge(x_1, \ldots, x_{2n-3})$. Now use the Wang sequence (Exercise 23).

Exercises on obstruction theory.
In the following exercises on obstruction theory, we assume that $F \to E \to B$ is a fibration where $\pi_0(B) = 0$ and $\pi_1(B)$ acts trivially on the cohomology of the fiber.

28. **Euler classes and Euler numbers.** Given $F \to E \xrightarrow{\pi} B$, assume that $\pi_i(F) = 0$ for $i < n - 1$. Denote by $B^{(k)}$ the k–skeleton of B (always assuming that B is a CW complex).
Show that
(a) there exists a cross-section $B^{(n-1)} \xrightarrow{\sigma} E$ (recall that a cross-section σ of $E \to B$ is a map $\sigma : B \to E$ such that $\pi\sigma =$ identity. Think of this as $\sigma(x)$ is contained in F_x, the fiber over x, for all $x \in B$).
(b) Any two sections $B^{n-2} \underset{\sigma_2}{\overset{\sigma_1}{\longrightarrow}} E$ are homotopic.

(c) The obstruction to finding $B^{(n)} \xrightarrow{\sigma} E$ is given by a cohomology class $c(E) \in H^n(B, H_{n-1}(F))$.

Remark. (a), (b), (c) imply that in a spherical fibration $S^{n-1} \to E \to B$, there exists a unique class $e(E) \in H^n(B, \mathbb{Z})$, the Euler class, such that $S^{n-1} \to E \to B$ has a section over $B^{(n)} \Leftrightarrow e(E) = 0$. In case B is an oriented n-manifold, we may define the *Euler number* of the spherical fibration by

$$e(E) \; = < e(E), B > .$$

In this case $S^{n-1} \to E \to B$ has a section if and only if the Euler number e(E) is zero.

In this exercise, use that for each n-cell e^n. on B,

$$E|_{e^n} \sim e^n \times F.$$

29. Let B be a compact, oriented n-manifold and $\mathbb{R}^n \to E_0 \to B$ an oriented vector bundle (Exercise: Give the definition). Choose a metric in $\mathbb{R}^n \to E_0 \to B$ and let $S^{n-1} \to E \to B$ be the unit sphere bundle.

    Let e(E) be the Euler number as defined in the previous exercise, so that we have a nowhere zero section of $\mathbb{R}^n \to E_0 \to B \Leftrightarrow e(E) = 0$.

    (a) Verify that to compute e(E), you do the following:

    Find a nonzero section $B^{(n-1)} \xrightarrow{\sigma} E^*$ ($E^* = E -$ zero section) (here we are assuming that B has been triangulated). For each n-cell e^n_α, we have $E|e^n_\alpha \approx e^n_\alpha \times R^n$ and so σ gives $\partial e^n_\alpha \xrightarrow{\sigma} \mathbb{R}^n - \{0\}$, or equivalently

    $$\partial e^n_\alpha \xrightarrow{\sigma_\alpha} S^{n-1}$$

    then

    $$e(E) = \Sigma_\alpha \, \text{degree}(\sigma_\alpha).$$

    Why is it sufficient to use <u>any</u> section σ?

    (b) Consider the universal line bundle $\mathbb{C} \longrightarrow E \xrightarrow{\pi} \mathbb{P}^1$ where $\pi^{-1}[z_0, z_1]$ is the complex line through (z_0, z_1) in $\mathbb{C}^2$. Using the Euclidean metric on $\mathbb{C}^2$, the associated sphere bundle is the Hopf fibration

    $$S^1 \longrightarrow S^3 \longrightarrow \mathbb{P}^1.$$

Over $\mathbb{P}^1 - \{\infty\}$, where ∞ is the point with homogeneous coordinates [0, 1]), choose the cross-section

$$\sigma([z_0, z_1]) = (1, z_1/z_0).$$

Using this, show that

$$e(E) = -1.$$

30. **Definition of Chern classes using obstruction theory.**

(a) Recall the Stiefel manifold $S(n-k+1, n)$ of $n-k+1$ frames in $\mathbb{C}^n$. We computed (cf. Exercise 16)

$$\pi_i(S(n-k+1, n) = 0, \quad i < 2k-1$$

$$\pi_{2k-1}(S(n-k+1, n)) \cong \mathbb{Z}.$$

Let $\mathbb{C}^n \to E \to B$ be a complex vector bundle over a CW complex B. Using Exercise (28), deduce that there is a class

$$c_k(E) \in H^{2k}(B, \mathbb{Z})$$

such that $\mathbb{C}^n \to E \to B^{(2k)}$ has a field of $n-k+1$ frames $\Leftrightarrow$ $c_k(E) = 0$. Show that if we have a diagram of mappings

$$\begin{array}{ccc} E & \xrightarrow[F]{} & E' \\ \downarrow & & \downarrow \\ B & \xrightarrow[f]{} & B' \end{array}$$

with F a linear isomorphism on the fibers, then $f^* c_k(E') = c_k(E)$. The classes $c_0(E) = 1,\ c_1(E),\ \ldots,\ c_n(E)$ are the Chern classes of $\mathbb{C}^n \to E \to$

(b) Using Exercise 29(b), show that for the universal line bundle $\mathbb{C} \to E \to \mathbb{P}^1$ the class $c_1(E)$ is equal to minus the standard generator in $H^2(\mathbb{P}^1, \mathbb{Z})$ coming from the complex orientation of $\mathbb{P}^1$.

(c) Let $\mathbb{C} \to E_n \to \mathbb{P}^n$ be the universal line bundle over $\mathbb{P}^n(\pi^{-1}(z_0, \ldots, z_n) =$ line $(tz_0, \ldots, tz_n)$ in $\mathbb{C}^{n+1}$). Using Exercise 30(a) above, show that

$$c_1(E_n) = -g$$

where $g \in H^2(\mathbb{P}^n, \mathbb{Z})$ is the generator that restricts to the natural generator on $\mathbb{P}^1$.

Remark. Given a vector bundle $\mathbb{C}^n \to E \to B$, let $P(E)$ be the bundle of lines in E; thus, the fiber $P(E)_x = P(E_x)$ is the set of lines through the origin in $E_x \cong \mathbb{C}^n$. Representing points in $P(E)$ as pairs (x, ξ), where ξ is a line in E_x, there is a line bundle $\mathbb{C} \to L(E) \xrightarrow{\pi} P(E)$ where $\pi^{-1}(x, \xi) = \{\text{line } \xi\} \subset E_x$. Consider the fibration $\mathbb{P}^{n-1} \to P(E) \to B$. The first Chern class $c_1(L(E)) \in H^2(P(E), \mathbb{Z})$

restricts to -{ generator of $H^2(\mathbb{P}^{n-1}; \mathbb{Z})$} by Exercise 30(a) and (c). From this, we deduce that $\mathbb{P}^{n-1} \to P(E) \to B$ satisfies the condition in Exercise 25, and so for $\zeta = -c_1(L(E))$,

$$H^*(P(E)) \cong H^*(B)[\xi]/(\zeta^n - \Sigma_{q=1}^n (-1)^{n-q} c_q(E)\zeta^{n-q}).$$

The classes $c_q(E) \in H^{2q}(B)$ are the same Chern classes as those defined by obstruction theory (this is not trivial). The Chern classes are the fundamental invariants for measuring the nontriviality of a vector bundle $\mathbb{C}^n \to E \to B$.

31. **Hopf theorem on singularities of a vector field.** Given a vector field $\theta = \Sigma_i \theta_i \, \partial/\partial x_i$ in a neighborhood of the origin in $\mathbb{R}^n$ and having an isolated zero at $x = 0$, there is associated an integer, the *index* $\text{ind}_0(,,)$ of θ at $x = 0$, defined as follows:

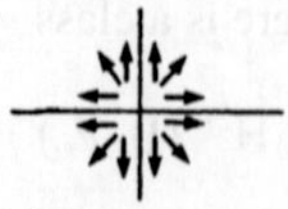
(index +1)

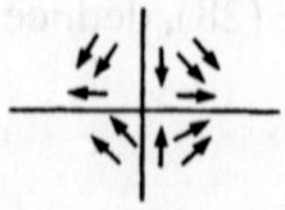
(index -1)

In a small sphere $\|x\| = \varepsilon$, the function $\tilde{\theta}(x) = (\theta_1, \ldots, \theta_n)$ is nonzero, and so $f(x) = \tilde{\theta}(x)/\|\tilde{\theta}(x)\|$ gives a map

$$S^{n-1} \xrightarrow{f} S^{n-1}.$$

Then, $\text{ind}_0(\theta) = \text{degree}(f)$.
Let B be a compact, oriented n-manifold and θ a vector field having only isolated zeros.

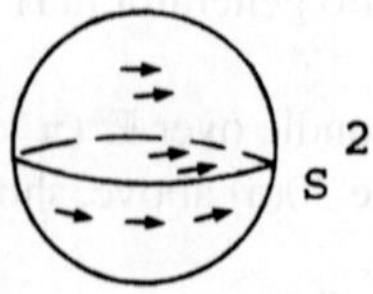

θ is rotational vector field

We may view θ as a section of the tangent bundle $T \to B$. Prove the formula

$$e(T) = \Sigma_{\theta(x)=0} \, \text{ind}_x(\theta)$$

Remark. You may assume that B has a CW decomposition with only cells of dimension $\leq$ n (why?) and that θ has zeros only at interior points of the n-cells. Now try to apply Exercise 29(a).

This proves that $\Sigma_{\theta(x)=0}\ \mathrm{ind}_X(\theta)$ is independent of the vector field θ. To actually calculate $\Sigma_{\theta(x)=0}\ \mathrm{ind}_X(\theta)$ for a particular vector field θ, you make a smooth triangulation of B and use the second barycentric subdivision to find a particularly nice vector field. There is a beautiful intuitive discussion of this on pp. 201–203 of [25]. For this vector field θ, you find the formula

$$\Sigma_{\theta(x)=0}\ \mathrm{ind}_X(\theta) = \Sigma(-1)^p\{\#\text{ of p-cells in cell decomposition of B}\}$$

Putting these formulas together and using Exercise 21 gives the final result

$$e(T) = \chi\ (B) = \Sigma_{\theta(x)=0}\ \mathrm{ind}_X(\theta)$$

which is a famous theorem of Poincaré-Hopf.

32. **Two lemmas.**
   (a) Let $X' = X \cup_f e^{n+1}$ be obtained by attaching an $(n+1)$-cell to X by $\partial e^{n+1} \xrightarrow{f} X$. Using the definition, show that the obstruction to extending $X \xrightarrow{id} X$ to $X' \to X$ is $[f] \in \pi_n(X)$.
   (b) Given $A \subset X$, a cohomology class $\alpha \in H^q(X)$, and a cocycle $\tilde{\alpha} \in \tilde{Z}^q(A)$ which represents the restriction $\alpha|A$, show that there is a cocycle $\tilde{\alpha}' \in \tilde{Z}^q(X)$ such that $\tilde{\alpha}'|A = \tilde{\alpha}$. (Here we are using cellular cochains.)

33. **Cellular approximation theorem.** Suppose that X, Y are CW complexes and $X \xrightarrow{f} Y$ is a continuous map. Show that f is homotopic to a map $X \xrightarrow{g} Y$ which is *cellular* in the sense that

$$g(X^{(n)}) \subset Y^{(n)}$$

for the respective n-skeleton. (Hint: For simplicity, assume first that $\dim X < \infty$ and suppose by induction that we have a map $X \xrightarrow{f_n} Y$ homotopic to f with $f_n(X^{(n)}) \subset Y^{(n)}$.
Show now that we have

$$\pi_{n+1}(Y^{(n+1)}, Y^{(n)}) \longrightarrow \pi_{n+1}(Y, Y^{(n)}) \longrightarrow 0$$

and that $\pi_{n+1}(Y^{(n+1)}, Y^{(n)})$ is a free abelian group on the $n+1$ cells in X (use relative Hurewicz together with $H_q(Y, Y^{(n)}) = 0$ for $0 \le q \le n$).
Now let e^{n+1} be an $n+1$ cell of X. Then, $f_n\colon (e^{n+1}, \partial e^{n+1}) \to (Y, Y^{(n)})$ is an element in $\pi_{n+1}(Y, Y^{(n)})$. By the above remark, f_n may be deformed into $\tilde{f}_n\colon (e^{n+1}, \partial e^{n+1}) \to (Y^{(n+1)}, Y^{(n)})$. We will be done if we show that, during the homotopy $f_n \sim \tilde{f}_n$, the map may be kept constant on ∂e^{n+1} (why is this necessary?). Thus, we must prove the

Lemma. *Given* $A \subset B \subset Y$, *if* $(e^{n+1}, S^n) \xrightarrow{\alpha} (Y, A)$ *is homotopic to* $(e^{n+1}, S^n) \xrightarrow{\beta} (B, A)$, *then* α *is homotopic to* $(e^{n+1}, S^n) \xrightarrow{\gamma} (B, A)$ *where all maps in the homotopy are constant on* $S^n = \partial e^{n+1}$.

Proof. Given

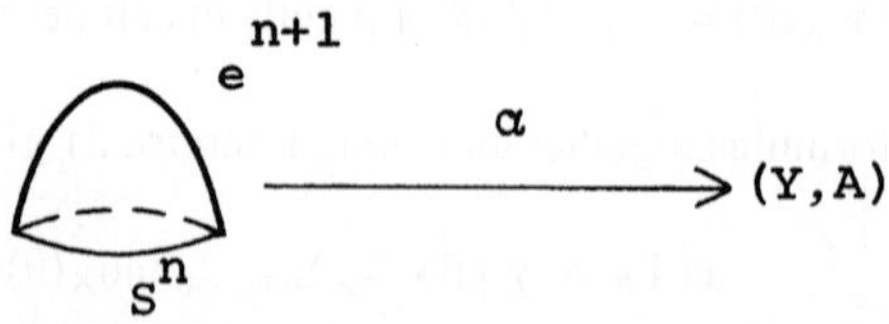

and a homotopy α_t of α with

$$\alpha_0 = \alpha$$
$$\alpha_1 = \beta$$
$$\alpha_t(S^n) \subset A,$$

put a "collar" around $S^n \hookrightarrow S^n \times I$ and define γ_t by the picture

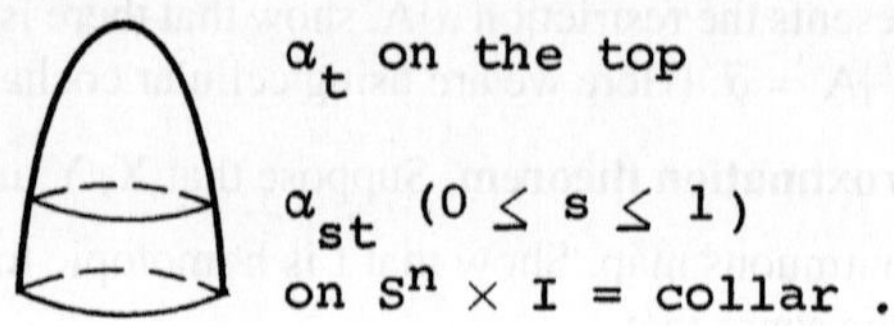

Proceeding inductively over the (n + 1)-cells, we now have

$$f_{n+1}: X^{(n+1)} \longrightarrow Y^{(n+1)}$$
$$f_{n+1}|X^{(n)} = f_n$$
$$f_{n+1} \sim f \text{ on } X^{(n+1)}$$

Apply the homotopy extension property to extend f_{n+1} to all of X. ■

What modification is necessary in case $\dim X = \infty$?

34. **Whitehead products.** (Preliminary exercises on wedges and products of spheres.)

(a) Let $S^p \vee S^q \subset S^p \times S^q$ be the usual embedding. Considering the n-sphere as

$$S^n = I^n/\partial I^n,$$

show that:

(i) The quotient $S^p \times S^q / S^p \vee S^q$ is homeomorphic to S^{p+q}.

(ii) $H^\ell(S^p \times S^q,\ S^p \vee S^q;\ G) \cong H^\ell(S^{p+q};\ G)$, $\ell > 0$ for any coefficient group G.

(Hint: For (i), $S^p \times S^q/S^p \vee S^q$

$$= (I^p/\partial I^p \times I^q/\partial I^q)/(I^p/\partial I^p \times *) \cup (* \times I^q/\partial I^q)$$
$$\sim I^{p+q}/\partial I^{q+q} \quad \text{(why?)}.$$

For (ii) use the previous exercise about excision.)

(Note: This show that $S^p \times S^q = (S^p \vee S^q) \cup_f e^{p+q}$ where $f\colon S^{p+q-1} \to S^p \vee S^q$, for some map f representing an element $[f] \in \pi_{p+q-1}(S^p \vee S^q)$.)

(b) Recall the definition of the *Whitehead product*: Given $S^p \vee S^q \subset S^p \times S^q$, the obstruction to extending the identity map $S^p \vee S^q \xrightarrow{i} S^p \vee S^q$ to $S^p \times S^q \to S^p \vee S^q$ is a cohomology class „(i) $\in H^{p+q}(S^p \times S^q, S^p \vee S^q; \pi_{p+q-1}(S^p \vee S^q)) \cong \pi_{p+q-1}(S^p \vee S^q)$. By definition the Whitehead product

$$W_{p,q} \in \pi_{p+q-1}(S^p \vee S^q)$$

corresponds to $\theta(i)$ under this isomorphism.

Using Exercise 32(a) and Part (i) above, we have $W_{p,q}$ is the class of the attaching map $\partial e^{p+q} \xrightarrow{f} S^p \vee S^q$ used in constructing $S^p \times S^q$ by attaching a cell to $S^p \vee S^q$.

By considering the cohomology rings of $S^p \times S^q$ and $S^p \vee S^q \vee S^{p+q}$, show that $W_{p,q} \neq 0$.

If $\alpha \in \pi_p(X)$ and $\beta \in \pi_q(X)$ are represented by maps $\tilde{\alpha}\colon S^p \to X$ and $\tilde{\beta}\colon S^q \to X$, then the Whitehead product

$$[\alpha, \beta] \in \pi_{p+q-1}(X)$$

is represented by the composite mapping

$$S^{p+q-1} \xrightarrow[f]{} S^p \vee S^q \xrightarrow[\tilde{\alpha}\vee\tilde{\beta}]{} X$$

where f is the above attaching map. Thus

$$[\alpha, \beta] = (\alpha \vee \beta)_*(W_{p,q})$$

where $W_{p,q} \in \pi_{p+q-1}(S^p \vee S^q)$ was defined above. Show that the Whitehead product satisfies the relations:

$$[\alpha, \beta] = (-1)^{p,q}[\beta, a] \qquad \text{(symmetry)}$$
$$[\alpha, [\beta, \gamma]] = [[\alpha, \beta], \gamma] + (-1)^{pq}[\beta, [\alpha, \gamma]] \qquad \text{(Jacobi identity)},$$

where $\alpha \in \pi_p(X)$ and $\beta \in \pi_q(X)$.

35. **Transgression.** (The general case discussed here is not often used. The most important case is the one dealt with in Exercise 36.) Let $F \to E \xrightarrow{\pi} B$ be

a fibration with $\pi_0(B) = \{0\}$, $\pi_1(B) = \{0\}$, $\pi_1(F) = \{0\}$, so that the Serre spectral sequence is applicable. Consider the cohomology sequences (with $q > 0$)

$$\begin{array}{ccccccc} H^q(E) & \longrightarrow & H^q(F) & \xrightarrow[\delta]{} & H^{q+1}(E,F) & \longrightarrow & H^{q+1}(E) \\ & & & & \uparrow{\scriptstyle \pi^*} & & \uparrow{\scriptstyle \pi^*} \\ & & & & H^{q+1}(B,x_0) & \xrightarrow[\cong]{} & H^{q+1}(B) \end{array}$$

The *transgression* τ is the map from the subgroup

$$T^q(F) = \delta^{-1}\pi^* H^{q+1}(B) \subset H^q(F)$$

to the quotient group

$$G^{q+1}(F) = H^{q+1}(B)/\ker \pi^* \nu$$

defined by

$$\tau\,(\delta^{-1}\pi^*\alpha) = \alpha.$$

This map is of fundamental importance in the cohomology theory of bundles. Using the proof of the Serre spectral sequence, show that for $q \geq 1$

(i) $T^q(F) \approx E^{0,q}_{q+1} \subset E^{0,q}_2 \approx H^q(F)$.
(ii) $G^{q+1}(F) \cong E^{q+1,0}_{q+1}$ is a quotient of $E^{q+1,0}_2 \cong H^{q+1}(B)$.
(iii) $\tau = d_{q+1}\colon E^{0,q}_{q+1} \to E^{q+1,0}_{q+1}$.

(Hint: This requires that you look carefully into the construction of the spectral sequence. To get some idea, try the case $q = 1$.

$q = 1$: To prove (i) and (ii) in this case, you must show that $H^2(B) \xrightarrow{\pi^*} H^2(E,F)$ is an isomorphism. This follows from $H_2(E,F) \approx H_2(B)$ (use the relative Hurewicz isomorphism and $\pi_i(E,F) \approx \pi_i(B)$) and the fact that $H^2(B)$, $H^2(E,F)$ are torsion free (why? use $\pi_1(B) = 0 = \pi_1(E,F)$.)
The fact that $\tau = d_2$ in this case follows from the definition of d_2 as a coboundary in an exact cohomology sequence of pairs $(B^{(n)}, B^{(n-2)})$ where $B^{(q)} = \pi^{-1}(X^{(q)})$.

36. **Some examples of transgression.** (This exercise depends on Exercise 35, which you should in any case read over carefully first; cf. also the discussion at the beginning of Chap. 12.)

(a) Suppose that $F \to E \to B$ is a fibration and $H^q(E) = 0$ for $q > 0$ (e.g., if E is contractible). Show that

$$\{H^i(F) = 0 \text{ for } 0 < i < q\}$$

$$\Leftrightarrow \{H^j(B) = 0 \text{ for } 0 < j < q + 1\}$$

and that the transgression is an isomorphism

$$\tau: H^q(F) \xrightarrow{\sim} H^{q+1}(B).$$

(Hint: It will suffice to assume that $H^i(F) = 0 = H^j(B)$ for $0 < i < q$, $0 < j < q + 1$ and show that

(i) $E_2^{0,q} = \ldots = E_{q+1}^{0,q}$
(ii) $E_2^{q+1,0} = \ldots = E_{q+1}^{q+1,0}$
(iii) $d_{q+1} : E_{q+1}^{0,q} \tilde{\to} E_{q=1}^{q+1,0}$ is an isomorphism.

The E_2-term of the Serre spectral sequence looks like

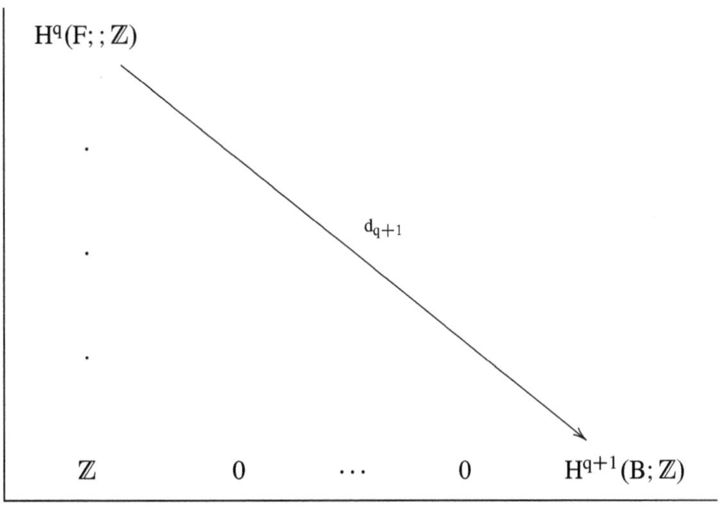

From this, it follows that $d_r = 0$ on $E_2^{0,q}$ for $2 \leq r \leq q$, and this gives (i) and (ii).

Now then $d_{q+1} : E_{q+1}^{0,q} \to E_{q+1}^{q+1,0}$ must be an isomorphism, since otherwise we would get a nontrivial element in $E_\infty^{0,q}$ or $E_\infty^{q+1,0}$, but $E_\infty = 0$ since $H^*(E) = 0$.)

(b) Let

$$\begin{array}{ccc} K(\pi,\ n-1) & \to & \mathscr{P} \\ & & \downarrow \\ & & K(\pi,\ n) \end{array}$$

be the path fibration ($n \geq 2$). Show that in the Serre spectral sequence,

$$E_2^{0,\,n-1} = \ldots = E_n^{0,\,n-1} \cong H^{n-1}(K(\pi,\ n-1))$$

$$E_2^{n,\,0} = \ldots = E_n^{n,\,0} \cong H^{n}(K(\pi,\ n-1))$$

$$E_n^{0,\,n-1} \xrightarrow{d_n} E_n^{n,\,0} \text{ is an isomorphism.}$$

(Hint: Use part (a) of this exercise above.)

(c) Show that if we have a map of fiber spaces

$$\begin{array}{ccc} F' & \longrightarrow & F \\ \downarrow & & \downarrow \\ E' & \xrightarrow[\tilde{f}]{} & E \\ \downarrow & & \downarrow \\ X' & \xrightarrow[f]{} & X \end{array}$$

then transgression is natural. (This follows either from the definition or from the spectral sequence interpretation, both of which are given in Exercise 35.)

37. **Chern classes and transgression.** Recall the Stiefel manifold $S(n, N)$ of n-frames in $\mathbb{C}^N$ and that

$$\pi_q(S(n, N)) = 0, \quad 0 < q < 2N - 2n + 1.$$

Using the inclusions $\mathbb{C}^N \subset \mathbb{C}^{N+1}$, we have

$$S(n, N) \subset S(n, N + 1)$$

and if we let

$$S(n) = \text{n-frames in } \mathbb{C}^\infty$$

be the infinite Stiefel manifold, then

(i) $\pi_q(S(n)) = 0$ for $q > 0$

(If you don't like "∞" here, take $S(n, N)$ for N arbitrarily large.) Next, recall the Grassmann manifold $G(n)$ of n-planes in $\mathbb{C}^\infty$ and the fibration

$$U(n) \to S(n) \to G(n)$$

where S(n is the Stiefel manifold of unitary n-frames in $\mathbb{C}^\infty$ and the map $S(n) \to G(n)$ assigns to each frame the subspace that it spans. Recall that

(ii) $H^*(G(n)) = \mathbb{C}[x_1, x_2, \ldots, x_n]$ is a polynomial algebra with generators

$$x_q \in H^{2q}(G(n)).$$

Finally, recall that for the unitary group, the cohomology

(iii) $H^*(U(n)) = \wedge(y_1, \ldots, y_n)$

is an exterior algebra with generators

$$y_q \in H^{2q-1}(U(n)).$$

Prove that each y_q transgresses and

$$\tau(y_q) \equiv x_q \mod (x_1, \ldots, x_{q-1})$$

(Hint: From equation (i) it follows that $H^q(S(n)) = 0$ for $q > 0$. The E_2 of the spectral sequence looks like

$$\begin{array}{c|cccccccc}
y_3 & \mathbb{Z} & & & & & & & \\
y_2y_1 & \mathbb{Z} & & & & & & & \\
y_2 & \mathbb{Z} & & & & & & & \\
 & 0 & & & & & & & \\
y_1 & \mathbb{Z} & & & & & & & \\
 & \mathbb{Z} & 0 & \mathbb{Z} & 0 & \mathbb{Z}\oplus\mathbb{Z} & 0 & \mathbb{Z}\oplus\mathbb{Z}\oplus\mathbb{Z} \\
\hline
 & & & x_1 & & \{x_1^2, x_2\} & & \{x_1^3, x_1x_2.x_3\}
\end{array}$$

with $y_q \in H^{2q-1}(U(n))$ and $x_q \in H^{2q}(G(n))$.
Now argue as follows:

$d_2y_1 = x_1$ since $H^2(S(n)) = 0$

$d_2y_2 = 0 = d_3y_3$ because of the $0'$s in the spectral sequence

$\Rightarrow y_2$ is transgressive and $d_4y_2 \equiv x_2 \mod (x_1^2)$

$d_2y_3 = 0$ since $d_2y_1y_2x_1 = x_1^2y_2 + x_1x_2y_2 \neq 0$

and $d_3y_3 = d_4y_3 = d_5y_3 = 0$ because of $0'$s in the spectral sequence

$\Rightarrow y_3$ is transgressive and $d_6y_3 \equiv x_3 \mod (x_1x_2)$, etc.)

Remark. It can be proved that $x_q \in H^{2q}(G(n))$ is the qth Chern class of the universal vector bundle

$$\begin{array}{ccc}
\mathbb{C}^n & \longrightarrow & E \\
 & & \downarrow \\
 & & G(n)
\end{array}$$

Taking into account Exercise 30 above, this gives a third definition of Chern classes (these definitions are all the same up to sign!) Roughly speaking, these various definitions have the following advantages:

$$\left\{\begin{array}{l}\text{Definition 1 using}\\ \text{obstruction theory}\end{array}\right\} \longleftrightarrow \left\{\begin{array}{l}\text{Useful for "classical"}\\ \text{algebraic topology, such as}\\ \text{existence of vector fields, etc.}\end{array}\right\}$$

$$\left\{\begin{array}{l}\text{Definition 2 using}\\ \text{the projective bundle}\\ \mathbb{P}(\mathrm{E})\end{array}\right\} \longleftrightarrow \left\{\begin{array}{l}\text{Quickest definition and the}\\ \text{easiest one to prove the}\\ \text{duality theorem, also it is}\\ \text{useful in algebraic geometry}\end{array}\right\}$$

$$\left\{\begin{array}{l}\text{Defn. 3 using}\\ \text{transgression and}\\ \text{cohomology of the}\\ \text{Grassmannians}\end{array}\right\} \longleftrightarrow \left\{\begin{array}{l}\text{This definition is useful in}\\ \text{algebraic geometry and also}\\ \text{is the definition which shows}\\ \text{how to compute Chern classes}\\ \text{using curvature via deRham's}\\ \text{theorem}\end{array}\right\}$$

38. **Cohomology of** $K(\mathbb{Z}, 3)$**.** Prove that

$$H^q(K(\mathbb{Z}, 3); \mathbb{Z}) = \begin{cases} \mathbb{Z} & q = 0,\ 3 \\ \mathbb{Z}/2\mathbb{Z} & q = 6 \\ 0 & \text{otherwise for } 0 \le q \le 6. \end{cases}$$

(Hint: Use the fibration

$$\begin{array}{ccc} K(\mathbb{Z}, 2) & \to & \mathcal{P} \\ & & \downarrow \\ & & K(\mathbb{Z}, 3) \end{array}$$

together with the Serre spectral sequence and the fact that

$$H^q(\mathcal{P}) = 0 \quad (q > 0) \quad (\text{why?})$$

$$\begin{aligned} H^*(K(\mathbb{Z}, 2)) &= H^*(\mathbb{C}P^\infty) \\ &= \mathbb{Z}[x], \quad x \in H^2(K(\mathbb{Z}, 2)). \end{aligned}$$

The E_2 term has $H^*(K(\mathbb{Z}, 2))$, which is a polynomial algebra on a generator x in degree 2, up the left-hand side and $H^*(K(\mathbb{Z}, 3))$ along the bottom row. Thus, in the range, we are considering $d_2 = 0$ and d_3 looks like

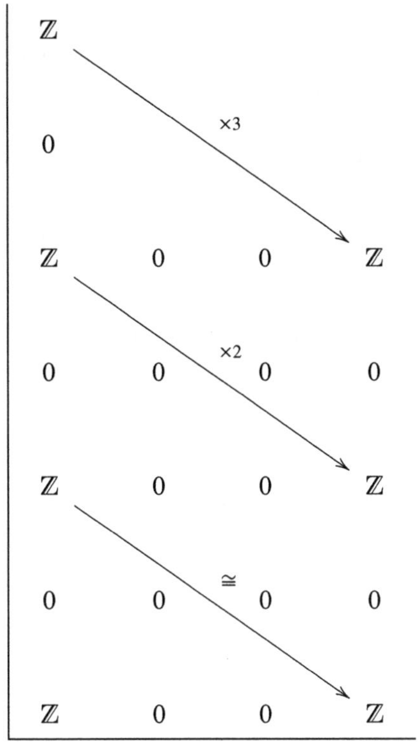

where $y \in H^3(K(\mathbb{Z}, 3))$ is a generator, by Hurewicz.
Then, show that $d_2 = 0$ for low values and

$d_3x = y$ (why?)
$d_3x^2 = 2xy$ (why?)
$d_3y = 0$ (obvious).
$\Rightarrow E_3^{3,2}/d_3E_3^{0,4} \cong \mathbb{Z}/2\mathbb{Z}$ with generator xy.

But $E_4^{3,2} = 0$ (why? use $E_\infty^{3,2} = 0$).

$$\Rightarrow \ker\{E_3^{3,2} \xrightarrow{d_3} E_3^{6,0}\} = \text{i.m}\{E_3^{0,4} \xrightarrow{d_3} E_3^{3,2}\}$$

$\Rightarrow E_3^{6,0} \cong \mathbb{Z}/2\mathbb{Z}$ (to prove this you must show that $H^4(K(\mathbb{Z}, 3)) = 0$

$\Rightarrow H^6(K(\mathbb{Z}, 3)) \cong \mathbb{Z}/2\mathbb{Z}$.

39. **Homotopy groups of spheres.** Prove that

$$\pi_4(S^2) \cong \mathbb{Z}/2\mathbb{Z}.$$

(Hint: To do this, use the Postnikov tower (= P. T.) for the 2-sphere S^2. Letting $(S^2)_n$ be the nth stage of the P. T., recall that there is a map $S^2 \xrightarrow{f_n} (S^2)_n$ such that $\pi_i(S^2) \cong \pi_i((S^2)_n) \quad i \leq n$ and

$$H_i(S^2) \underset{(f_n)_*}{\xrightarrow{\sim}} H_i((S^2)_n) \qquad 0 \leq i \leq n+1.$$

Step one: $(S^2)_2 = K(\mathbb{Z}, 2) = \mathbb{C}P^\infty$ since $\pi_2(S^2) = \mathbb{Z}$.
Step two: There is a fibration

$$\begin{array}{ccc} K(\mathbb{Z},3) & \to & (S^2)_3 \\ & & \downarrow \\ & & K(\mathbb{Z},2) = (S^2)_2 \end{array}$$

since $\pi_3(S^2) \cong \mathbb{Z}$. Moreover, $H^4((S^2)_3) = 0$. Look in the spectral sequence of this fibering, whose E_2 is, using the previous exercise,

$\mathbb{Z}/2\mathbb{Z}$				
0				
0				
$\mathbb{Z}$				
0				
0				
$\mathbb{Z}$	0	$\mathbb{Z}$	0	$\mathbb{Z}$

Show then that $d_4 y = x^2$, and from this, deduce that for $0 \leq q \leq 6$

$$H^q((S^2)_3) = \begin{cases} \mathbb{Z} & q = 0,\ 2 \\ \mathbb{Z}/2\mathbb{Z} & q = 6 \\ 0 & \text{otherwise} \end{cases}$$

(Note: You must also use that $d_4(yx) = x^3$.)

Step three: Let $\pi = \pi_4(S^2)$. Then, the next stage of the P.T. is

$$\begin{array}{ccc} K(\pi, 4) & \rightarrow & (S^2)_4 \\ & & \downarrow \\ & & (S^2)_3 \end{array}$$

and $H^q((S^2)_4) = 0$ for $2 < q \leq 5$. From this, deduce that

(a) π is a finite abelian group (it might be 0)
(b) $H^4(K(\pi, 4); \mathbb{Z}) = 0$ and
(c) $H^5(K(\pi, 4); \mathbb{Z}) \cong \pi$.

Thus, the E_2 term of the spectral sequence is

π						
0						
0						
0						
0						
$\mathbb{Z}$	0	$\mathbb{Z}$	0	0	0	$\mathbb{Z}/2\mathbb{Z}$

Deduce from $E_\infty^{s,t} = 0$, $3 \leq s + t \leq 5$ that

$$d_6 \colon E_6^{0,5} \longrightarrow E_6^{6,0}$$

is an isomorphism. Thus, $\pi \cong \mathbb{Z}/2\mathbb{Z}$.

40. **Classification theorem for complex vector bundles.** Let X be a CW complex of dimension n and $\mathrm{Vect}^r(X)$ be the isomorphism classes of r-plane bundles $\mathbb{C}^r \to E \to X$. Denote by $G(r, N)$ the Grassmannian of r-planes in $\mathbb{C}^N$ and by

$$\begin{array}{ccc} \mathbb{C}^r & \longrightarrow & U_r \\ & & \downarrow \pi \\ & & G(r, N) \end{array}$$

the universal vector bundle (thus, π^{-1} of an r-plane is the same r-plane considered as a vector space). Show that the assignment

$$\begin{aligned} [X, G(r, N)] &\to \mathrm{Vect}^r(X) \\ f &\mapsto f^* U_r \end{aligned}$$

is a bijection for $2N - 2r + 1 > n$.

Proof. Recall the fibration $U_r \to S(r, N) \to G(r, N)$, where $S(r, N)$ is the Stiefel manifold of r-frames in $\mathbb{C}^N$ and also that

$$\begin{aligned} \pi_i(S(r, N)) &= 0 \quad \text{for } i < 2N - 2r + 1 \\ \Rightarrow \pi_i(S(r, N)) &= 0 \quad \text{for } i < n + 1 \end{aligned}$$

under the conditions of the theorem. Let's first show that every vector bundle $\mathbb{C}^r \to E \to X$ is of the form $f^* U_r$ for some map $X \to G(r, N)$. Letting $X^{(k)}$ be the k-skeleton, suppose we have

$$\begin{aligned} &X^{(k)} \xrightarrow{f^{(k)}} G(r, N) \\ &f^{(k)*} U_r \cong E | X^{(k)}. \end{aligned}$$

Now the restriction of E to each $(k + 1)$-cell e_{k+1} is trivial (why?), so that we have

$$\begin{aligned} &E | e_{k+1} \cong e_{k+1} \times \mathbb{C}^r \\ &\partial e_{k+1} \xrightarrow{f^{(k)}} G(r, N) \\ (*) \qquad &\Rightarrow f^{(k)*} E | \partial e_{k+1} \cong \partial e_{k+1} \times \mathbb{C}^r. \end{aligned}$$

This gives us a lifting of $\partial e_{k+1} \xrightarrow{f^{(k)}} G(r, N)$ to

$$\begin{array}{ccc} & & S(r, N) \\ & \overset{g}{\nearrow} & \downarrow \\ \partial e_{k+1} & \xrightarrow{f^{(k)}} & G(r, N) \end{array}$$

by taking the r-frame to be the coordinate frame in $\mathbb{C}^r$ under the isomorphism $(*)$.

Since $\pi_k(S(r, N)) = 0$, the map g extends over e_{k+1}, and in this way, we may extend $f^{(k)}$ over the $(k+1)$ -skeleton (why?). Thus,

$$[X, G(r, N)] \longrightarrow \mathrm{Vect}^r(X)$$

is onto, and now a relative version of the same argument shows that it is one-to-one.) ∎

Remark. A special case of this theorem is the isomorphism

$$\mathrm{Vect}^r(S^n) \cong \pi_n(G(r, N)) \cong \pi_n(BU_r)$$
$$(N \text{ large relative to } r, n)$$

We call two vector bundles E, E' on a space X *stably isomorphic* if

$$E \oplus t^\ell \approx E' \oplus t^{\ell'}$$

where t^ℓ is the trivial bundle of rank ℓ. The equivalence classes of stable vector bundles will be denoted by K(X), and then the above theorem gives

$$K(X) \cong [X, BU].$$

Returning to the n-sphere, we have then

$$K(S^n) \cong \pi_n(BU),$$

and so Bott periodicity computes the stable vector bundles over spheres.

41. **Axioms for homology.** Suppose we are given a covariant functor

$$X \to \mathscr{H}_*(X) = \oplus_{p \geq 0} \mathscr{H}_p(X)$$

from finite CW complexes to abelian groups such that

(i) $X \overset{f}{\to} Y$ induces $\mathscr{H}_*(X) \overset{f*}{\to} \mathscr{H}_*(Y)$ which depends only on the homotopy class of f.

(ii) $\mathscr{H}_*(\text{pt}) = \begin{cases} \mathbb{Z} & * = 0 \\ 0 & * > 0 \end{cases}$

(iii) Given $X \subset Y$, if we define

$$\mathscr{H}_*(Y, X) \cong \mathscr{H}_*(Y/X)$$

(this property forces excision to be true), then the exact homology sequence for a pair holds.

Show that $\mathscr{H}_*(X) = H_*(X; Z)$ is ordinary homology.

(Hint: By (ii), this is true if $\dim X = 0$. Suppose by induction it is true when $\dim X < n$. If D^n is the n-disk with $\partial D^n = S^{n-1}$, then $\mathscr{H}_*(D^n) = \{\begin{matrix} \mathbb{Z}, & * = 0 \\ 0, & n > 0 \end{matrix}\}$ by (i) and $\mathscr{H}_*(S^{n-1}) \cong \{\begin{matrix} \mathbb{Z}, & * = 0, n-1 \\ 0, & \text{otherwise} \end{matrix}\}$ by induction. The exact homology sequence then gives

$$(*) \qquad \mathscr{H}_*(D^n, \partial D^n) \cong H_*(D^n, \partial D^n).$$

Suppose now that $X = Y \cup_f e^n$ where $\dim Y \leq n - 1$. Then, we have

$$\begin{array}{ccccccc} H_p(Y) & \longrightarrow & H_p(X) & \longrightarrow & H_p(X, Y) & \xrightarrow{\partial} & H_{p-1}(Y) \\ & & & & \wr\wr & & \\ \mathscr{H}_p(Y) & \longrightarrow & \mathscr{H}_p(X) & \longrightarrow & \mathscr{H}_p(X, Y) & \longrightarrow & \mathscr{H}_{p-1}(Y) \end{array}$$

where the isomorphism $\mathscr{H}_p(X, Y) \approx H_p(X, Y)$ follows by $(*)$ above.

Taking $p = n$, we get a map $\mathscr{H}_n(X) \to H_n(X)$ which is an $\cong$. For $p = n-1$, we have

$$H_{n-1}(X) \cong H_{n-1}(Y)/H_n(X, Y)$$

and

$$\mathscr{H}_{n-1}(X) \cong \mathscr{H}_{n-1}(Y)/\mathscr{H}_n(X, Y).$$

This gives a map $\mathscr{H}_{n-1}(X) \to H_{n-1}(X)$ which is an isomorphism. The remaining $\mathscr{H}_p$, H_p for $p < n - 1$ are already isomorphic (why?).

Now complete the proof by induction on the number of n-cells in X.

Remark. The analogous theorem for cohomology is also true with the same proof. In general, an *extraordinary cohomology theory* is a contravariant functor

$$X \to K^*(X)$$

from spaces X to abelian groups which satisfy the cohomology axioms corresponding to (i) and (iii), but not necessarily the axiom specifying the value on the one-point space. Using Bott periodicity and denoting by $S^n X$ the n-fold suspension of X, one obtains K-theory with

$$K^n(X) \underset{\text{defn.}}{=} K(S^n X)$$

is an extraordinary cohomology theory.

42. **Little problems on filtrations.**

(i) Show that if $A = A_0 \supset A_1 \supset \ldots \supset A_m = \{0\}$ and $B = B_0 \supset B_1 \supset \ldots \supset B_n = \{0\}$ are filtered abelian groups and if $f: A \to B$ is a filtration preserving homomorphism inducing isomorphisms $A_p/A_{p+1} \xrightarrow{\sim} B_p/B_{p+1}$, then f itself is an isomorphism.

(ii) Using (i) show that if a map of Serre spectral sequences induces an isomorphism on E_∞, then the abutments are isomorphic.

(iii) As an example of the failure of the converse of (ii), let X be a CW complex with base point $x_0 \in X$. Consider the trivial fibrations $X \to x_0$ and $X \xrightarrow{\text{id}} X$. The identity map $X \to X$ induces a map of these fibrations which is an isomorphism on the abutment of the spectral sequences (which is $H^*(X)$ in both cases). Show from the definitions that this map is not an isomorphism on E_∞ unless $H^*(X) = 0$ for $* > 0$.

43. **The comparison theorem for spectral sequences.** Let $E^r_{p,q}$ $E'^r_{p,q}$ be two homology spectral sequences (same as cohomology spectral sequences with upper and lower indices interchanged and with arrows reversed; also the proof of the Serre spectral sequence for homology is the same as for cohomology). Suppose that we have a map

$$E^r_{p,q} \to E'^r_{p,q} \quad \text{(for all r and the map commutes with } d_r\text{'s)}$$

such that

$$E^2_{0,q} \cong E'^2_{0,q}$$

and

$$E^\infty_{p,q} \cong E'^\infty_{p,q} \quad \text{for all } p,\ q \geq 0.$$

Show that $E^2_{p,0} \cong E'^2_{p,0}$ for all $p \geq 0$.

Remark. This may be interpreted as saying that if we have a map between fiber spaces

$$\begin{array}{ccc} F & \longrightarrow & F' \\ \downarrow & & \downarrow \\ E & \longrightarrow & E' \\ \downarrow & & \downarrow \\ B & \longrightarrow & B' \end{array}$$

inducing isomorphisms on $H_*(F) \xrightarrow{\sim} H_*(F')$ and $H_*(E) \xrightarrow{\sim} H_*(E')$, then $H_*(B) \to H_*(B')$ is also an isomorphism (compare the previous exercise).

Step one: Show that

$$E^2_{p,0} \cong E'^2_{p,0} \text{ for a fixed p} \Rightarrow$$

$$E^2_{p,q} \cong E'^2_{p,q} \text{ for all q.}$$

Step two: Prove that $E^2_{0,0} \cong E'^2_{0,0}$ and

$$E^2_{1,0} \cong E'^2_{1,0}.$$

Step three: Check that $E^3_{0,1} \xrightarrow{\sim} E'^3_{0,1}$ (use $E^\infty_{0,1}$);

$$E^2_{0,1} \to E'^2_{0,1} \text{ and } E^3_{2,0} \xrightarrow{\sim} E'^3_{2,0} \quad (\text{use } E^\infty_{2,0}).$$

Use the exact sequence

$$0 \to E^3_{2,0} \to E^2_{2,0} \to \operatorname{Ker}\{E^2_{0,1} \to E^3_{0,1}\} \to 0$$

to conclude that

$$E^2_{2,0} \xrightarrow{\sim} E'^2_{2,0}.$$

Step four: For general p, we attack $E^r_{p,0}$ by descending induction on $r \geq 2$ until we find $E^2_{p,0} \xrightarrow{\sim} E'^2_{p,0}$.

First $E^{p+1}_{p,0} \xrightarrow{\sim} E'^{p+1}_{p,0}$ (why? use $E^\infty_{p,0}$) and $E^{p+1}_{0,p-1} \xrightarrow{\sim} E'^{p+1}_{0,p-1}$ (why?). Using $E^2_{0,p-1} \xrightarrow{\sim} E'^2_{0,p-1}$ and ascending induction on r, show that $E^p_{0,p-1} \xrightarrow{\sim} E'^p_{0,p-1}$. Conclude that $E^p_{p,0} \xrightarrow{\sim} E'^p_{p,0}$.

Step five: Using ascending induction on k, prove that

$$E^k_{p-r,s} \xrightarrow{\sim} E^k_{p-r,s} \text{ for } 2 \leq k \leq r,\ s \geq 0.$$

In particular,

$$E^{r}_{p-r,r-1} \overset{\sim}{\to} E'^{r}_{p-r,r-1}.$$

Using this and descending induction on $k \geq 1$, prove that $E^{r+k}_{p-r,r-1} \overset{\sim}{\to} E'^{r+k}_{p-r,r-1}$. In particular,

$$E^{r+1}_{p-r,r-1} \overset{\sim}{\to} E'^{r+1}_{p-r,r-1}.$$

Step six: Show that image $\{E^{r}_{p,0} \to E^{r}_{p-1,r-1}\} \overset{\sim}{\to}$ image $\{E'^{r}_{p,0} \to E'^{r}_{p-1,r-1}\}$ using $E^{r+1}_{p-r,r-1} \overset{\sim}{\to} E'^{r+1}_{p-r,r-1}$ and the fact that ker $\left\{E^{r}_{p-r,r-1} \to E^{r}_{p-2r,2r-2}\right\}$ agree with ker $\left\{E'^{r}_{p-r,r-1} \to E'^{r}_{p-2r,2r-2}\right\}$.

Conclude that $E^{r}_{p,0} \overset{\sim}{\to} E'^{r}_{p,0}$ using $E^{r+1}_{p,0} \overset{\sim}{\to} E'^{r+1}_{p,0}$.

44. **Homotopy of wedges of 2-spheres.** Let $X = S^2_1 \vee \ldots \vee S^2_m$ be a wedge of 2-spheres. Then the second homotopy group

$$\pi_2(X) \cong \oplus_{i=1}^{m} \pi_2(S^2_i) \cong \mathbb{Z}^m,$$

and there are Whitehead products

$$\pi_{ij} = [S^2_i, S^2_j] \in \pi_3(X).$$

Show that these products for $i \leq j$ give a free $\mathbb{Z}$-basis for $\pi_3(X)$. Thus,

$$\pi_3\left(\vee_{i=1}^{m} S^2_i\right) \cong \mathbb{Z}^{m(m+1)/2}.$$

(Hint: This problem is a good exercise in Postnikov towers. Since $\pi_2(X) \cong \oplus_{i=1}^{m} \pi_2(S^2_i)$, the P.T. for X begins with

$$X_2 = \Pi_{i=1}^{m} K(\mathbb{Z}, 2).$$

The next step is

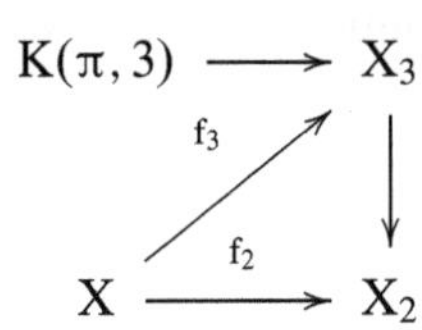

We want to determine what π is. From the exact cohomology sequence of (X_2, X), we have

$$\begin{array}{ccccccc} H^3(X,\pi) & \to & H^4(X_2,X;\pi) & \to & H^4(X_2,\pi) & \to & H^4(X,\pi) \\ \| & & & & & & \| \\ 0 & & & & & & 0 \end{array}$$

Since $\pi_i(X_2, X) = 0$ for $0 \le i \le 3$ (why?), the pair (X_2, X) begins with cells in dimension 4, and so

$$\begin{aligned} H^4(X_2, X; \pi) &\cong \mathrm{Hom}(H_3(X_2, X), \pi) \\ &\cong \mathrm{Hom}(\pi, \pi) \text{ (why?)}. \end{aligned}$$

Thus, we have

$$0 \longrightarrow \mathrm{Hom}(\pi, \pi) \longrightarrow H^4(X_2, \pi) \longrightarrow 0.$$

Since $H_*(X_2; \mathbb{Z})$ is free, we obtain then

$$0 \longrightarrow \mathrm{Hom}(\pi, \pi) \xrightarrow{\sim} \mathrm{Hom}(H_4(X_2), \pi) \longrightarrow 0.$$

From this, we see that π is free abelian of finite rank and

$$\mathrm{Hom}(\pi; \mathbb{Z}) \cong H^4(X_2; \mathbb{Z}) \cong \mathbb{Z}^{m(m+1)/2}.$$

Thus, π is a free abelian group of rank $m(m + 1)/2$. It remains to identify $\pi \cong \mathbb{Z}^{m(m+1)/2}$ as having a basis given by the Whitehead products $[S_i^2, S_j^2]$ $(i \le j)$. To do this, replace $K(\mathbb{Z}, 2)$ by $\mathbb{C}P^2$ in the P.T. for X. This is okay since $\pi_i(\mathbb{C}P^2) \cong \pi_i(K(\mathbb{Z}, 2))$ for $0 \le i \le 4$. Thus, we have

$$\begin{array}{ccc} K(\pi,3) & \longrightarrow & X_3 \\ & \nearrow & \downarrow \\ \vee S^2 & \longrightarrow & \prod \mathbb{C}P^2 \end{array}$$

($\vee\mathbb{P}^1$ is the 2-skeleton of $\Pi\, \mathbb{C}P^2$).
The k-invariant is by definition the primary obstruction to finding a section to the inclusion

$$\vee\mathbb{P}^1 \hookrightarrow \Pi\, \mathbb{C}P^2.$$

The primary obstruction is a class in

$$H^4\, (\Pi\, \mathbb{C}P^2, \pi_3(\vee\mathbb{P}^1)).$$

Recall that the Whitehead product of the generators of $\pi_2(S^2_0)$ and $\pi_2(S^2_1)$ is the obstruction to extending the identity $S^2_0 \vee S^2_1 \to S^2_0 \vee S^2_1$ to a map

$$S^2 \times S^2 \to S^2 \vee S^2.$$

This essentially completes the desired identification, once we show that the usual Hopf map $S^3 \to S^2$ represents the Whitehead product of the generator of $\pi_2(S^2)$ with itself (why?) Consider the 4-skeleton of $\Pi\ \mathbb{C}P^2$.

References

1. Baumslag, G.: Lecture notes on nilpotent groups. In: CBMS Regional Conference Series in Mathematics, vol. 2. American Mathematical Society, Providence (2007)
2. Chen, K.T.: Iterated path integrals. Bull. Amer. Math. Soc. **83**, 831–879 (1977)
3. Deligne, P., Griffiths, P., Morgan, J., Sullivan, D.: Homotopy theory of compact Kahler manifolds. Inv. Math. **29**, 245–274 (1974)
4. Donaldson, S.: An application of gauge theory to four-dimensional topology. J. Differ. Geom. **18**, 279–315 (1983)
5. Freedman, M.: The topology of four-dimensional manifolds. J. Differ. Geom. **17**, 357–453 (1982)
6. Fukaya, K., Oh, Y.-G., Ohta, H., Ono, K.: Lagrangian intersection Floer theory. AMS/IP Studies in Advanced Math, vol. 46. American Mathematical Society and International Press, Providence (2009)
7. Greenberg, M.: Lectures on Algebraic Topology. W.A. Benjamin, New York (1977)
8. Griffiths, P., Harris, J.: Principles of Algebraic Geometry. Wiley Classic Series, Wiley, New York (1994)
9. Hu, S.T.: Homotopy Theory. Academic, New York (1959)
10. Kadeishvili, T.: Cohomology C_∞-algebra and rational homotopy type. Banach Center Publ. **85**, 225–240 (2009). Available at arXiv0811.1655 (Math.AT)
11. Kan, D.: A combinatorial definition of homotopy groups. Ann. Math. **67**, 282–312 (1958)
12. Kirby, R., Siebenmann, L.: Foundational essays on topological manifolds, smoothings, and triangulations. In: Annals of Mathematics Studies vol. 88. Princeton Univeristy Press, Princeton (1977)
13. M. Kontsevich. Homological algebra of mirror symmetry. Available at arXiv:alg-geom/9411018v1, 1994.
14. Mal'cev, A.: On a class of homogeneous spaces. Amer. Math. Soc. Transl. **9**, 276–307 (1962)
15. Manolescu, C.: Pin(2)-equivariant Seiberg-Witten Floer homology and the triangulation conjecture (2013). Available at arXiv:1303.2354
16. Markl, M., Shnider, S., Stasheff, J.: Operads in algebra, topology and physics. In: Mathematical Surveys and Monographs, vol. 96. American Mathematical Society, Providence (2007)
17. Milnor, J.: On spaces having the homotopy type of a CW complex. Trans. Amer. Math. Soc. **50**, 272–280 (1959)
18. Morgan, J.: The algebraic topology of smooth algebraic varieties. Publ. IHES **48**, 137–204 (1978)
19. Quillen, D.: Rational homotopy theory. Ann. Math. **90**, 205–295 (1969)

P. Griffiths and J. Morgan, *Rational Homotopy Theory and Differential Forms*,
Progress in Mathematics 16, DOI 10.1007/978-1-4614-8468-4,

© Springer Science+Business Media New York 2013

20. Seidel, P.: Fukaya Categories and Picard-Lefschez Theory. Zurich Lecture Series in Advanced Mathematics. European Mathematical Society, Zurich (2008)
21. Serre, J.-P.: Homologie singulière des espaces fibrés. Ann. Math. **54**, 435–505 (1951)
22. Siebenmann, L.: Are non-triangulable manifolds triangulable? In: Cantrell, J.C., Edwards, C.H. (eds.) Topology of Manifolds, pp. 77–84. Markham, Chicago (1970)
23. Spanier, E.: Algebraic Topology. Springer, New York (1966)
24. Stasheff, J.: H-Spaces from the Homotopy Point of View. Spring Lecture Notes in Mathematics, vol. 161. Springer, Basel (1970)
25. Steenrod, N.: The Topology of Fiber Bundles. Princeton University Press, Princeton (1951)
26. Sullivan, D.: Infinitesimal calculations in topology. Publ. IHES **47**, 269–331 (1977)
27. Whitney, H.: Geometric Integration Theory. Princeton Press, Princeton (1957)

Progress in Mathematics
Edited by J. Coates and S. Helgason
Progress in Physics
Edited by A. Jaffe and D. Ruelle

- A collection of research-oriented monographs, reports, notes arising from lectures or seminars
- Quickly published concurrent with research
- Easily accessible through international distribution facilities
- Reasonably priced
- Reporting research developments combining original results with an expository treatment of the particular subject area
- A contribution to the international scientific community: for colleagues and for graduate students who are seeking current information and directions in their graduate and post-graduate work

Manuscripts

Manuscripts should be no less than 100 and preferably no more than 500 pages in length.

They are reproduced by a photographic process and therefore must be typed with extreme care. Symbols not on the typewriter should be inserted by hand in indelible black ink. Corrections to the typescript should be made by pasting in the new text or painting out errors with white correction fluid.

The typescript is reduced slightly (75 %) in size during reproduction; best results will not be obtained unless the text on any one page is kept within the overall limit of $6 \times 9^{1/2}$ in (16×24 cm). On request, the publisher will supply special paper with the typing area outlined.

Manuscripts should be sent to the editors or directly to:
Birkhäuser Boston, Inc., 380 Green St., Cambridge, Mass. 02139

Progress in Mathematics
Already published

PM1 Quadratic Forms in Infinite-Dimensional Vector Spaces
Herbert Gross
ISBN 3-7643-1111-8, 431 pages, $ 22.00, paperback

PM2 Singularités des systèmes différentiels de Gauss-Manin
Frédéric Pham
ISBN 3-7643-3002-3, 339 pages, $ 18.00, paperback

PM3 Vector Bundles on Complex Projective Spaces
C. Okonek, M. Schneider, H. Spindler
ISBN 3-7643-3000-7, 389 pages, $ 20.00, paperback

PM4 Complex Approximation, Proceedings, Quebec, Canada, July 3–8, 1978
Edited by Bernard Aupetit
ISBN 3-7643-3004-X, 128 pages, $ 10.00, paperback

PM5 The Radon Transform
Sigurdur Helgason
ISBN 3-7643-3006-6, 202 pages, $ 14.00, paperback

PM6 The Weil Representation, Maslov Index and Theta Series
Gérard Lion, Michèle Vergne
ISBN 3-7643-3007-4, 345 pages, $ 18.00, paperback

PM7 Vector Bundles and Differential Equations Proceedings, Nice, France, June 12–17, 1979
Edited by André Hirschowitz
ISBN 3-7643-3022-8, 255 pages, $ 14.00, paperback

PM8 Dynamical Systems, C.I.M.E. Lectures, Bressanone, Italy, June 1978
John Guckenheimer, Jürgen Moser, Sheldon E. Newhouse
ISBN 3-7643-3024-4, 300 pages, $ 14.00, paperback

PM9 Linear Algebraic Groups
T. A. Springer
ISBN 3-7643-3029-5, 304 pages, $ 18.00, hardcover

PM 10 Ergodic Theory and Dynamical Systems I
A. Katok
ISBN 3-7643-3036-8, 352 pages, $ 20.00, hardcover

PM11 Eighteenth Scandinavian Congress of Mathematicians, Aarhus, Denmark, 1980
Edited by Erik Balslev
ISBN 3-7643-3034-6, 528 pages, $ 26.00, hardcover

PM12 Séminaire de Théorie des Nombres, Paris 1979–80
Edited by Marie-José Bertin
ISBN 3-7643-3035-X, 408 pages, $ 22.00, hardcover

PM13 Topics in Harmonic Analysis on Homogeneous Spaces
Sigurdur Helgason
ISBN 3-7643-3051-1, 142 pages, $ 12.00, hardcover

PM14 Representations of Real Reductive Lie Groups
David A. Vogan, Jr.
ISBN 3-7643-3037-6, 771 pages, $ 35.00, hardcover

Progress in Physics
Already published

PPh1 Iterated Maps on the Interval as Dynamical Systems
Pierre Collet and Jean-Pierre Eckmann
ISBN 3-7643-3026-0, 256 pages, $ 16.00 hardcover

PPh2 Vortices and Monopoles, Structure of Static Gauge Theories
Arthur Jaffe and Clifford Taubes
ISBN 3-7643-3025-2, 275 pages, $ 16.00 hardcover

PPh3 Mathematics and Physics
Yu. I. Manin
ISBN 3-7643-3027-9,111 pages, $ 10.00 hardcover

MIX
Papier aus verantwortungsvollen Quellen
Paper from responsible sources
FSC® C105338

If you have any concerns about our products,
you can contact us on
ProductSafety@springernature.com

In case Publisher is established outside the EU,
the EU authorized representative is:
Springer Nature Customer Service Center GmbH
Europaplatz 3, 69115 Heidelberg, Germany

Printed by Libri Plureos GmbH
in Hamburg, Germany